ICE

High-tech on rails

Editors-in-chief:

Dipl.-Ing. Dipl.-Wirtsch.-Ing. Theo Rahn
President of the Bundesbahn-Zentralamt München

Professor Dr.-Ing. Hubert Hochbruck
Manager of the Forschungsgemeinschaft Rad/Schiene

Ministerialrat Friedrich W. Möller
Member of the Presidency of the Verkehrsforum Bahn e.V.

Executive editor:

Dipl.-Ing. Dietmar Lübke
Departmental President in the Bundesbahn-Zentralamt München

Hestra-Verlag

ISBN 3-7771-0231-8

Printed in West-Germany

Editor's foreword to the 2nd (revised) edition

The year 1985 saw the 150th birthday of the German Railways, even as the newly developed power units and centre trailers of the Intercity-Experimental (ICE/V) high-speed multiple unit were completed, put into operation, and subjected to thorough tests, culminating, on 26th November of the same year, in the setting up of a new German speed record and a world record for three-phase multiple units as the ICE/V reached 317 kmph on the Bielefeld–Essen line between Guetersloh and Neubeckum.

During the following months, the ICE/V undertook systematic trials. It proved its efficiency during innumerable test and demonstration runs. The German Federal Railway, universities, research establishments, and the German railway industry found their concept of a new generation of high-speed rolling stock borne out by the results of the test runs.

On 1st May 1988, the ICE/V achieved a new world speed record for rail vehicles at 406.9 kmph on the new line Wuerzburg–Fulda and became the world first train to exceed 400 kmph (250 mph).

The knowledge amassed from the trial and demonstration runs formed the basis for the planning and construction of production vehicles ordered by the German Federal Railway. These multiple units, with two power units and fourteen centre trailers, will enter high-speed service between Hamburg and Munich, and, at a later stage, also to Switzerland.
On the new lines, the InterCityExpress trains will run at 250 kmph, – on the new line Cologne–Frankfurt at 300 kmph – and on the upgraded lines at 200 kmph, considerably shortening the existing journey times.

Great comfort and short journey times will be the most attractive features of this new rail transport development.

Nor will the ICE high-speed transport systems stop at international frontiers.
Multi-system vehicles will be used to link European centres of commerce and industry; they will be able to run at high speeds on all European rail networks from the Netherlands to Italy and from France to Poland.

In the spring of 1989, the German Federal Government asked the German Federal Railway and the German railway industry to develop a multi-system variant of the ICE.
This train will contribute towards an even more closely-knit railway system in Europe.

This new edition of ICE HIGH-TECH ON RAILS describes the progress made from the test and demonstration train to the final product. In this new edition, the partners in the development of the ICE, encouraged by the great success of the 1st edition, outline the experiences and their practical application to the introduction of the high-speed rail service.

The editors

Contents

ICE

The Authors

Dipl.-Ing. (FH) Manfred Ackermann

Since 1983 Head of the sales department of high-voltage equipment at the AEG-Westinghouse Transport-Systeme GmbH. – Project leader for the AEG's share of the electrical equipment of the ICE-trailers. – 1952–1956 Apprenticeship for high-voltage current at AEG. – 1957–1960 Study of electrical engineering at the Beuth Technical College Berlin. – 1960–1968 Project engineer for high-voltage equipments and from 1968 to 1973 additional leading of high-voltage mounting. – 1973–1982 Assistant manager of the Head of the department of high-voltage equipment. – Address: AEG-Westinghouse Transport-Systeme GmbH, Hohenzollerndamm 150, D-1000 Berlin 33.

Prof. Dr.-Ing. Karlheinz Althammer

Director General of the German Federal Railways – Study of Mechanical Engineering at the Technical University München Subsequent to his service in the Bundesbahn-Direktion München, Scientific Assistant at the Technical University München; – Scientific Asisstant at the Hochleistungs-Schnellbahn-Studiengesellschaft (HSB); Head of a District Mechanical Office; – Head of Section at the Bundesbahn-Zentralamt München; during that period, Project Leader for the "Construction and Commissioning of the Roller Rig Test Facility München-Freimann" as well as for the "Test Line Rheine–Freren"; in between, Member of the Working Group "New Network Concept" at DB Headquarters; since 1973, Lecturer at the Technical University Braunschweig for "Guided Land Vehicles"; – since 1980, Director of the Bundesbahn-Versuchsanstalt München. – Address: Bundesbahn-Versuchsanstalt München, Völckerstrasse 5, D-8000 München 45.

Dipl.-Ing. Sven Bartels

Since 1984 Project leader for development of technology and mechanics, device-design at Dornier GmbH. – 1974–1978 Study of technology and mechanical engineering at the Technical University Stuttgart. – Manufactoring and design of equipment at IPA Stuttgart. – Examiner for metal-lines at the IHK Weingarten. – Address: Dornier GmbH, Abteilung MRG, Postfach 1420, D-7990 Friedrichshafen 1.

Dr.-Ing. Klaus Becker-Lindhorst

Department manager in the railway brake division of Knorr-Bremse AG, München. He studied general mechanical engineering at the "Technische Universität" in Braunschweig. 1969 to 1975 engaged on production engineering development with Siemens AG, München. In 1972, he acquired his doctorate at the institute for communication engineering of the "Technische Universität", Braunschweig. Since 1976 employed by Messrs. Knorr-Bremse AG, Moosacher Strasse 80, D-8000 München 40.

Dipl.-Ing. Heinz Böhm

Since 1983 Head of the department of locomotives inland at AEG-Westinghouse Transport-Systeme GmbH. – Study of electrical engineering at the Technical University Berlin. – 1970–1978 Project and development engineer at AEG-Bahntechnik. – 1978 Head of the department of development of electric drive systems of AC-vehicles. – Address: AEG-Westinghouse Transport-Systeme GmbH, Hohenzollerndamm 150, D-1000 Berlin 33.

Dipl.-Ing. Ulrich Budde

Development engineer with Krupp Industrietechnik GmbH in the product area "Railway Engineering", specialising in the areas of running techniques and testing. Studied mechanical engineering, specialists subject rail vehicles, at the Ruhr University in Bochum and the Technical University in Hannover. – Address: Krupp Industrietechnik GmbH, Helenenstrasse 149, D-4300 Essen 1.

Dipl.-Ing. Heinrich Gerdsmeier

Head of main department "Engineering – Rail Vehicles" of the product area "Railway Engineering" at Krupp Industrietechnik GmbH, Essen. Studied general mechanical engineering at the Technical College in Hannover. – Joined Krupp Industrietechnik in 1965 on design and development of rail vehicles. – Address: Krupp Industrietechnik GmbH, Helenenstrasse 149, D-4300 Essen 1.

Dipl.-Ing. Christian R. Günther, Ph. D. Director Engineering and Programs

MBB-Transport Technology, a Division of the Industrial Products Group – 1966 M. S. Technical University Darmstadt in the field of control systems and mechanical engineering – 1977 PH.D. University of California, Los Angeles in the field of control systems and systems engineering – 1967–1973 Systems Analyst for control systems and guided vehicles with Garrett AiResearch, Los Angeles. – since 1973 Project Engineer with MBB New Transport Systems, München – until 1983 Chief Engineer Research and Development Rolling Stock and Aircraft Components with MBB Helicopter and Transport Systems. – Address: Messerschmitt-Bölkow-Blohm GmbH, Industrial Products Group, MBB-Transportation Technology, P. O. Box 1353, D-8850 Donauwörth.

Dipl.-Ing. Werner Herbert

Head of a main department for development of automatic system, plants and motor drive equipment at Dornier System GmbH. – Study of mechanical engineering at the Technical University Darmstadt. – Address: Oberer Garwiedenweg 10, D-7778 Markdorf.

Dipl.-Ing. Heinz R. Kurz

Bundesbahndirektor. Section head in the Department for Research and Development Planning at the Bundesbahn-Zentralamt München. – Study of Mechanical Engineering at the Technical University Darmstadt. – Service in the Motive Power Department and in the Workshop Department of DB; subsequently, Scientific Assistant in the Mechanical Department of the BZA München. 1973–1977, seconded to work with the Deutsche Eisenbahn Consulting GmbH (head of a team attending the Wheel-on-Rail Research Programme of the Federal Ministry of Research and Technology). Subsequently, Head of Section responsible for vehicles and for track/vehicle dynamics in connection with the systems coordination of the Wheel-on-Rail Research Programme; since 1978, member of the Department for Research and Development Planning (study of systems approach in connection with the Wheel-on-Rail System; Project Leader for the ICE multiple unit). – Address: Bundesbahn-Zentralamt München, Arnulfstrasse 19, D-8000 München 2.

Dipl.-Ing. Walter Lößel

Senior Engineer Assistant head of Development Office in Traction Systems and Vehicles Subdivision of Siemens AG. 1955–1960: Studied electrical engineering at München Technical University. – Aug. 1960: Joined the Traction Department of Siemens AG, Erlangen as Project Engineer for electric locomotives and motor cars. Worked on the development, planning and construction of standard locomotives and power cars for the German Federal Railway and for a number of foreign railway authorities. – Sept. 1982: Appointed Head of the ICE Project for the Siemens section of delivery, as well as spokesman for the electrical equipment of the power units. – Address: Siemens AG, Traction Department E 41, Werner-von-Siemens-Strasse 50, D-8520 Erlangen.

Dipl.-Ing. Dietmar Lübke

Departmental President. Since 1980, Head of the Department for Research and Development Planning at the Bundesbahn-Zentralamt München. – Study of Mechanical Engineering at the Technical University München. – Following the second State Examination in March 1966, service in the Motive Power Department of the DB Regional Headquarters München; Scientific Assistant at the Bundesbahn-Zentralamt München. 1970–1973, service in the Federal Ministry of Transport, Division A. 1973–1974, Divisional Manager of the District Mechanical Office Rosenheim; subsequently, Head of Section 102, then Head of Section 107 in the Department for Research and Development Planning at the BZA München. Address: Bundesbahn-Zentralamt München, Arnulfstrasse 11, D-8000 München 2.

Dr.-Ing. Gerhard Pintag

Deutsche Eisenbahn-Consulting GmbH, Head of the Project Assistant Team acting on behalf of the German Federal Ministry for Research and Technology in the field of "Guided Ground Transport in Wheel-on-Rail Technology". – Studies of Electrical Engineering (telecommunications, data processing) dissertation on aerospatial technology at the Technical University of Darmstadt. – Address: Deutsche Eisenbahn-Consulting GmbH, Oskar-Sommer-Strasse 15, D-6000 Frankfurt/Main 70.

Dipl.-Ing. Andreas Schaefer-Enkeler

Head of the design of electrical locomotives and maglev-vehicles at Krauss-Maffei-Verkehrstechnik GmbH. – Study of mechanical engineering at the Technical University Aachen. – 1984 joined Krauss-Maffei AG as project engineer, section of traffic. – Since 1985 at department of design of electrical locomotives, since 1988 head of this department. – Address: Krauss-Maffei Verkehrstechnik GmbH, Krauss-Maffei-Strasse 2, D-8000 München 50.

Dipl.-Ing. Dieter Schmidt

Project leader for the electrical equipment of the ICE motor units. – Study of electrical engineering at the Technical University Karlsruhe. – Joined at the department of traffic at Brown Boverie & Cie AG at Mannheim in 1969. – Groupleader at the department of electrical locomotives and railcars at ABB Henschel Lokomotiven GmbH. – Address: ABB Henschel Lokomotiven GmbH, Abteilung Lok/BA 3, Neustadter Strasse 62, D-6800 Mannheim 31.

Dipl.-Ing. (FH) Lutz Schwendt

Chief Engineer Design and Development Locomotive End Product. – Graduate (mechanical engineering) of Technical College Coburg, Joined Thyssen Henschel 1968. – Address: Thyssen Industrie AG Henschel, Henschelplatz 1, D-3500 Kassel.

Dipl.-Ing. Wolfgang von Waldstätten

Since 1980 Technical manager at the Linke-Hoffmann-Busch GmbH. – 1952–1957 Study of mechanical engineering at the Technical University Munich. – Joined Linke-Hoffmann-Busch GmbH in 1957 on design and development of rail vehicles. – Since 1974 Head of the department of design and development of rail vehicles. – Address: Linke-Hoffmann-Busch GmbH, Postfach, D-3320 Salzgitter.

Dr.-Ing. E.h. Horst Weigelt

Since 1979 President of the Bundesbahndirektion Nürnberg. – Study of civil engineering at the Technical University Darmstadt. – 1955 Joined at DB-business. – 1957 Scientific assistant at the Technical University Berlin. – 1960 joined at the civil engineer's and operating department of the Bundesbahndirektion Hamburg. – 1963 Divisional manager of the Hamburg City-S-Bahn. – 1965 Head of the department of rapid transit at the Hamburger Verkehrsverbund. – 1970 Member of the managing administration of the ite-institute Hamburg. – 1974 Head of the department of new S-Bahn lines. – Address: Bundesbahndirektion Nürnberg, Sandstrasse 38, D-8500 Nürnberg.

Dipl.-Ing. Ulrich Westerkamp

Since 1977, Director of the Bundesbahn-Versuchsanstalt Minden (Westf.) – Study of Railway Mechanical Engineering at the Technical University Hannover. Following the Second State Examination, Section Head at the Workshop Bremen; Member of the Joint Coordinating Staff Group for the BZA Minden and München; Head of the District Mechanical Office Kassel; Head of Section for Mechanical Installations at the Regional Headquarters Hamburg. – Address: DB-Versuchsanstalt Minden, Pionierstrasse 10, D-4900 Minden (Westf.)

Dipl.-Ing. Josef Wiese

Deutsche Eisenbahn-Consulting GmbH, Member of the Department "Transport Technologies" and of the Project, Assistant Team acting on behalf of the German Federal Ministry for Research and Technology. – Studies of Mechanical Engineering (graduating in railway vehicles). – Graduation in 1974. – Civil servant of the German Federal Railway, now assigned to the Deutsche Eisenbahn-Consulting GmbH. – Address: Deutsche Eisenbahn-Consulting GmbH, Oskar-Sommer-Strasse 15, D-6000 Frankfurt/Main 70.

Prefaces

Dr. Heinz Riesenhuber
Federal Minister for Research and Technology

Because they are energy-conserving and largely independant of fuel oil, the railways are particularly environment-friendly. They are the transport medium which is eminently suitable for attaining important private and public enterprise objectives. Existing railway systems have suffered a considerable decline over the last few decades because of competition from road and air transport. Analysis of the background to this decline revealed that the outstanding criteria for the choice of a transport medium by passengers are superior comfort and higher speed. Railway technical innovations abroad have shown that this move away from the railway can be reversed by increased passenger appeal and competitiveness.

With the High-speed Rail Systems research programme, the Federal Ministry of Research and Technology (BMFT) promotes the development of new technologies, thereby contributing towards further progress in rail technology in the Federal Republic of Germany. It is the aim of this research to increase productivity and performance of track transport systems considerably and further to improve their environment friendliness, i.e. to adapt their transport capabilities to modern demands and to expand their share of the transport market.

Industry will thus benefit from a contribution towards its performance and competitiveness, not only on the domestic market, but also by holding its own in international competition through opening up and integrating the markets.

National and international investment in modern, high-speed railway systems are considerable. The resulting employment and economic perspectives are of great value. In the medium term, this applies to the tried-and-tested system of railways on the basis of the wheel-on-rail technology; but in the longer term, to the novel system of track transport, the high-speed maglev railway, which can supplement the conventional system in the higher speed range by using the power of the electro-magnet not only for propulsion, but also for contactless vehicle guidance.

The InterCityExpress (ICE) high-speed multiple units demonstrate clearly how productivity and performance of the wheel-on-rail system can be improved through far-reaching innovations.

The ICE production train adapted the basic concept of its forerunner, the IC Experimental (ICE/V). This development basis was abandoned only where commercial requirements, technological aspects derived from tests of the ICE/V, and economic considerations called for modifications.

The ICE represents an essential technological step forward for national railway technology in the Federal Republic of Germany. After the pioneer high-speed systems in Japan and in France, German railway technology, too, is now geared to provide proof of its high technological standards through its contribution in this field. The ICE is concrete evidence of this.

With the opening of the new and the upgraded lines, a high-speed rail network will be made available which will not only meet the demands of the market and the aim of cost effectiveness, but also be able to demonstrate the economic value of the railway for the benefit of the common good.

Dr. Heinz Riesenhuber

Hemjö Klein
Member of Board of German Federal Railway

The first century of the railway in Germany was characterised by its technical ascendancy, its monopoly on speed and the leading position in the transport market which resulted. As early as 1903, an electric railcar powered by three-phase current reached a top speed of 210 kmph. However, the invention of cars and aeroplanes followed. Consequent developments and investment in the railway's rivals have, however, led to serious market losses for the railway over the last 40 years. The percentage of the market held by rail passenger transport has sunk from approximately 40% in 1950 to less than 6% today.

Examples from France and Japan, however, demonstrate that this trend is reversible. The results of market research show that the railway has a good chance if it succeeds in using its particular advantages to offer attractive services. This means above all:

▷ proving the system's claim to being the fastest means of transport on land;
▷ offering more space than an aircraft and greater comfort than a bus;
▷ being more environment-friendly and conducive to communication than the car.

At present, the railway is fulfilling these requirements by:

▷ constructing and extending high-speed lines;
▷ developing and producing the high-speed InterCityExpress train (ICE);
▷ introducing the Inter-Regional (IR) system in addition to the IC system;
▷ connecting the provinces to the system by means of coordinated shuttle services;
▷ complementing the concept of long-distance transport by means of holiday express and motorail trains.

The heart of the product innovation "high-speed transport" is the ICExpress. Initially travelling at a speed of 250 kmph, it is planned to be in operation on the following routes from 1991 onwards:

▷ 1. Hamburg–Hanover–Fulda–Frankfurt/Main–Mannheim–Stuttgart–Munich;
and in the following year:
▷ 2. Hamburg–Hanover–Fulda–Nuremberg–Munich;
▷ 3. Hamburg–Hanover–Frankfurt/Main–Basle.

Further ICE lines will follow after construction of additional stretches, in particular between the Rhine/Ruhr and Rhine/Main conurbations.

The ICExpress' innovative high-speed technology will be thoroughly geared to the requirements of the passenger after a strict programme of tests for safety, reliability, environment friendliness, economicalness and suitability for series production which it underwent in its forerunner, the IC Experimental (ICE/V). The high point of the test programme was, without doubt, the world record run on 1st May 1988, on the stretch between Würzburg and Mottgers on the newly constructed Würzburg to Fulda line, when the ICE/V became the first rail vehicle to break the 400 kmph barrier, reaching a speed of 406.9 kmph.

The future of European transport is the railway. In January 1989, the 14 railways of the EC states, Austria and Switzerland drew up a plan to establish a European high-speed network by the year 2015. The total length of the network will be 30,000 km, and the total cost will be around DM 200 billion, DM 30 billion of which has already been allocated. The Federal Republic of Germany occupies a central position in the European network. Future European high-speed links will run on the Federal German Railway newly constructed and extended lines. The standards of the Federal German Railways in the field of high-speed travel will therefore play a central role in determining future possibilities for rail traffic. The Federal German Railways and the track vehicle industry are, with considerable support from the Federal Government, therefore working together on the future development of ICE technology to produce a multi-standard high-speed train "ICE/M", which will be able to operate on various currents throughout Europe. There is a good chance that the first ICE/M trainsets will be put into operation as planned in 1995, when the express route Paris–Brussels–Cologne–Amsterdam–London is opened, with a later connection to Frankfurt/Main and Dortmund.

We are confident that ICE technology will play a decisive role in the European high-speed network. For this reason, we trust that this book will not only be a contribution to the development of high-speed rail travel in Germany but that it will also show the way for the future of railways in Europe.

Hemjö Klein

Dipl.-Ing. Eberhard Kill
Member of the Board of Management of Siemens AG

The handing over of the first InterCityExpress (ICE) production series power-unit for the new high-speed multiple units of the German Federal Railway represents a further milestone towards rail transport providing competitive, fast connections across Germany and our neighbouring countries as part of a grand European high-speed transport concept.

Investments by the Federal German government have smoothed the way, and the railway construction sites in Germany are encouraging evidence that rail has a great future, especially in high-speed passenger, piggyback, and container trains, as well as in multi-mode transport involving road and air transport. This is of special significance in view of our already overcrowded roads and airspace.

Other countries, too, are making increasing efforts to restructure their rail networks. In Europe, extending as far south as the Iberian peninsula, high-speed connections scale on a large are under construction. The USA, Canada, Australia and Korea are planning major projects. For German industry, the railway sector is a highly important field of activity. There are more than 50,000 people employed in the fields of modern rapid transit, urban and underground railways as well as long-distance transport with its special requirements for light-weight construction and high speeds. Apart from the number of employees and a share in the export market of more than 50%, this activity in the rail technology of tomorrow is one of the stimulants for a multitude of new developments, such as light-weight construction of vehicle bodies and high-tech electronics with its increasing role in operation and automation functions. Increased space requirement in the machine room for additional components of high-speed technology, and for diagnostic tasks linked with the need for compactness, makes exacting demands on the skills of power-unit designers. Weight ratios of less than 0.5 t/m can only be achieved through the intensive use of electronic design aids based on finite-element methods.

Today's power-unit propulsion equipment uses three-phase AC systems with a variable-voltage link. New power-semiconductors, together with their associated cooling technologies, determine the design of the power-circuits.

The ICE instrument and control systems make maximum use of microprocessor technology. Compact control and diagnostic systems have thus been used extensively for the first time.

When the first ICE train shortly enters service on the new high-speed routes, it will prove that the political and financial commitment, the magnificent engineering effort, and interdisciplinary planning have all been worthwhile. An outstanding product has been created as a result of exemplary cooperation between the railway and industry. The international successes in this field of railway technology show that this effort has also greatly benefited the German economy and industrial employment.

Only those products that are used and acknowledged in the home market, and thus prove themselves, can form the basis for success in the export market. To achieve this, close, effective cooperation between operator, developer, and the production team is of the greatest importance. The German economy is grateful to the railway, the Federal Ministry of Research and Technology and especially to the Ministry of Transport for their active cooperation.

The ICE train is the hallmark of a high-speed railway with passenger appeal. The high-speed performance of the prototype train on its excellent permanent way and with its superb overhead line infrastructure made the experts sit up and take notice. The railway industry extends its best wishes to the ICE train for a lasting success in Germany, in Europe, and in the world-wide adoption of the new technologies that it uses. Thanks to the close cooperation between all the partners; the Bonn ministries, and the German Federal Railway, operational trials and entry into service on the high-speed lines are to be implemented shortly.

We wish this new political and technical development in transportation every success.

E. Kill

Dipl.-Ing. Eberhard Kill

Companies involved in the development and construction of the InterCityExpress

ABB HENSCHEL AG
Neustadter Straße 62
6800 Mannheim 31

AEG Westinghouse Transport-Systeme GmbH
Nonnendammallee 15–21
1000 Berlin 20
Mainzer Landstraße 349–351
6000 Frankfurt am Main 1

ANT Nachrichtentechnik GmbH
Gerberstraße 33
7150 Backnang

Bergische Stahl-Industrie –BSI– Verkehrstechnik
Papenberger Straße 38
5630 Remscheid 1

Dornier GmbH Deutsche Aerospace
An der Bundesstraße 31
7997 Immenstaad/Bodensee

Duewag Aktiengesellschaft
Duisburger Straße 145
4150 Krefeld 11

Flachglas AG
Auf der Reihe 2
4650 Gelsenkirchen

GEZ Gesellschaft für elektrische Zugausrüstung mbH
Flinschstraße 20
6000 Frankfurt am Main 60

Hagenuk GmbH
Westring 431
2300 Kiel

Konsortium Elektro- und Klimatechnik, KEK, ICE-Mittelwagen
Mitglieder des KEK:

Geschäfts- und Ansprechstelle:
AEG Westinghouse
Transport-Systeme GmbH
Mainzer Landstraße 349–351
6000 Frankfurt am Main 1

ABB HENSCHEL AG
Neustadter Straße 62
6800 Mannheim 1

AEG Westinghouse
Transport-Systeme GmbH (A 34)
Nonnendammallee 15–21
1000 Berlin 20

GEZ
Gesellschaft für elektrische
Zugausrüstung mbH
Flinschstraße 20
6000 Frankfurt am Main 60

Hagenuk GmbH
Westring 431
2300 Kiel

Krupp
Maschinentechnik GmbH
Helenenstraße 149
4300 Essen 1

Luwa Fahrzeugklimatechnik GmbH
Hanauer Landstraße 200
6000 Frankfurt am Main 1

PINTSCH-BAMAG
Antriebs- und Verkehrstechnik GmbH
Hünxer Straße 149
4220 Dinslaken

Schaltbau Gesellschaft mbH
Klausenburger Straße 6
8000 München 80

Knorr-Bremse AG
Moosacher Straße 80
8000 München 40

Krauss-Maffei Verkehrstechnik GmbH
Krauss-Maffei-Straße 2
8000 München 50

Krupp Maschinentechnik GmbH
Helenenstraße 149
4300 Essen 1

Linke-Hofmann-Busch Waggon-Fahrzeug-Maschinen GmbH
Postfach 41 11 60
3320 Salzgitter 41

Luwa Fahrzeugklimatechnik GmbH
Hanauer Landstraße 200
6000 Frankfurt am Main 1

MAN Technologie AG
Dachauer Straße 667
8000 München 50

Messerschmitt-Bölkow-Blohm GmbH
Postfach 13 53
8850 Donauwörth

PINTSCH-BAMAG Antriebs- und Verkehrstechnik GmbH
Hünxer Straße 149
4220 Dinslaken

Schaltbau Gesellschaft mbH
Klausenburger Straße 6
8000 München 80

Siemens AG Verkehrstechnik
Werner-von-Siemens-Straße 50
8520 Erlangen

Standard Elektrik Lorenz Aktiengesellschaft Unternehmensbereich Bahnen
Lorenzstraße 10
7000 Stuttgart 40

Thyssen Guss AG Komponenten und Systeme
Am Thyssenhaus 3
4300 Essen 1

Horst Weigelt

The history of high-speed trains on German railways

The railway – faster than the snipe

When the steam locomotive revolutionized surface transport, man was able to enter a range of unaccustomed speed, triggering both fear and pessimism, enthusiasm and optimism in others.

Two documents have come down to us from the early days of British railways, which illustrate by facts and reports the aspects of speed in 1829/30. The British railway pioneer Nicolas Wood in his Practical Textbook of Railway Matters of 1839 recorded the results of the Rainhill races from 6th to 8th October 1829 to an accuracy of 1 second [1], according to which the Rainhill victor, the locomotive "Rocket", achieved an average speed of 34.5 kmph on the 1.5 m measured track on its fastest run. The maximum recorded speed was 29⅓ mph, or 46.84 kmph.

How these unaccustomed speeds on the iron rails were experienced we learn from a letter by the young actress Frances Anne Kemble written on 25th August 1830, shortly before the railway line between Liverpool and Manchester was opened [2]. Together with her famous father, the actor Charles Kemble, she had been invited to a trial run, and was permitted to occupy what to her "appeared to be the place of the highest honour, next to George Stephenson".

Miss Kemble then reports on the return trip to Liverpool: "After the locomotive had taken on water... we departed at the locomotive's highest speed, 35 miles per hour – faster than a bird flies (we made the experiment with a snipe)... you have no idea of the feeling as we cut through the air. Nevertheless, the motion is as smooth as possible. I could have read and written on the train. – I rose, removed my hat, and drank in the air before me. The wind was strong, or it might have been our flying against it, I could not keep my eyes open".

Fig. 1: The "Adler" train: 1935 reproduction, leaving the Nuremberg-East Exhibition Station in the summer of 1985

Both documents, the matter-of-fact test report and the romanticizing letter, describe the level of speed already reached at that time and thus the starting basis for a sequence of trials, records, and product innovations, which can be described as the history of high-speed travel. This contribution deals mainly with the era up to the Second World War, because postwar development has already been described in more recent literature [12–15] and is also still contemporary.

From the passenger train to the fast train

How fast, then, were the first German railway trains? On the Ludwigs Eisenbahn between Nuremberg and Fuerth the 6–9-coach trains started service in 1835 at journey times of 12 to 15 minutes on a line of about 6 km length (Fig. 1). This corresponded to a commercial speed of 24 to 30 kmph. This engineer officer v. Berger recorded 6 minutes as fastest ride of the light engine, a top speed of 60 kmph [3]. The early German railways, opened only a few years later, operated within the same speed range. The passenger train between Leipzig and Dresden, for instance, ran at a commercial speed of 33 kmph in 1840 [4].

Even so, passenger trains were basically operated as through trains, because initially there were only a few intermediate stops, none at all between Nuremberg and Fuerth, and only eight on the 62 km line between Munich and Augsburg, compared with the 18 intermediate stations including S-Bahn today.

Pressure for further intermediate stations or halts for passenger transport gradually increased. If the maximum commercial speed of the Prussian passenger trains calling at every station is given as almost unchanged as 32 to 33 kmph from 1858 to 1879 [4], the reason for this may be due to the fact that improved locomotive performance was probably offset by more frequent stops.

After already 15 years of railway development in Germany a new type of train was introduced, whose speed was appreciably higher than that of the other trains: on 1st May 1851 the first Schnellzug (fast train) in Germany entered service between Berlin and Cologne. The new type of train was not characterized by dedicated locomotives and coaches; its distinguishing feature was a commercial speed increased by more than 25 %; this jump in speed was achieved by omission of intermediate stops at rural and small-town stations, which produced considerable gains in time in view of the prevailing low starting accelerations and decelerations. The commercial speed of the first fast train between Berlin and Cologne jumped from 30 to 38 kmph. On 15th November 1852 the Berlin–Cologne "courier train" reached a commercial speed as high as 41.4 kmph [5].

The new type of "Schnellzug" rapidly spread under this and other publicity-oriented designations such as Eilzug, Kurierzug (courier train), or Postschnellzug (fast mail train), obviously because these faster trains met the urgent demands of the travelling public. In 1855 as much as 11 % of all passenger- and freight trains in Prussia were already fast trains [5]. The commercial speed of the fastest trains gradually increased to 47 kmph by 1879 [4].

Those speeds would certainly not justify the description of "high-speed transport" in these days. Nevertheless, the "Schnellzug" undoubtedly represents the beginning of the history of high-speed transport, because for the first time a network of long-distance trains was systematically built up, offering up to 50 % higher commercial speeds than the passenger trains of the basic network. It was decided by the International Timetable Conference at Interlaken in 1889 to designate all fast-running trains with supplements fares as "fast trains". But already a more far-reaching product innovation was approaching.

Fig. 2: "Phoenix" was built in 1863 and remained in service until 1903. It was intended for fast train service and designed for a maximum speed of 90 kmph. During trial runs it is said to have reached 120 kmph.

The "D-Zug", a quality jump in speed, service, and comfort

On 1st May 1892, more than 40 years after the first fast train ran in Prussia, a new type of long-distance train, the express train, made its debut. This was a milestone in the history of high-speed transport, because a higher service speed was achieved not only through timetabling measures; in fact, for the first time a carefully planned means for long-distance travel was made available to the public; it evolved from the improvement and of the technology all the components of the operating system of the railway.

The characteristic features were the vestibule coaches with side corridor (Fig. 3) and the connections between them protected against the weather by gangway bellows. The official designation as "D-Zug" (express train) was derived from these "Durchgangswagen" (coaches with gangway connections),

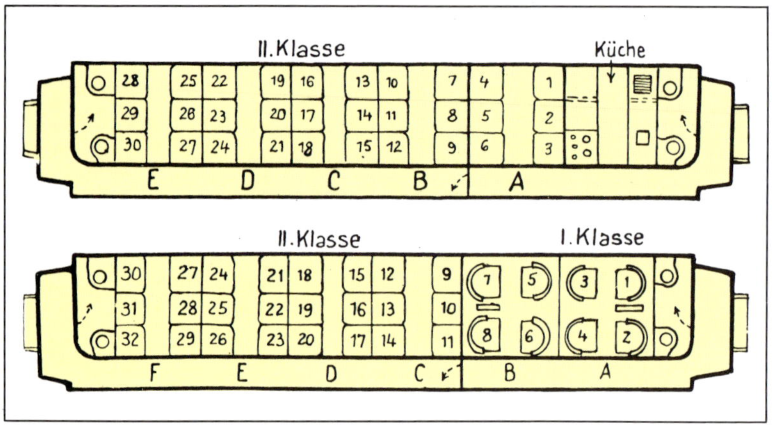

Fig. 3: Plans of the first vestibule coaches in Germany

and made a similar impression of progress and comfort on the contemporary travelling public to that of the IC85 on a higher level ninety years later.

The express train did not appear overnight. The origins of the vestibule coach with gangways were based on suggestions by Heusinger von Waldegg; since 1883 the legendary luxury train, the "Orient Express", had convincingly demonstrated the advantages of gangways between the coaches, and the attractions of comfortable coaches. For the first time unusual comfort and service was offered in a type of long-distance train dedicated to the upperclass travelling public on important main lines. The 1st Class featured wide swivel chairs which, however, gave rise to complaints as they tended to rock; they were later withdrawn. In the 2nd Class the new style compartments had only 3 seats per row, an unusually generous seating arrangement. The service, too, was innovative. The very first train included a kitchen leased to station restaurant owners; the gangway connections between the coaches made in-seat catering for the passengers possible. In-train catering brought about a further cut in journey times, through the elimination of meal stops of at least 25 minutes at major railway junctions with efficiently run station restaurants [10, 1932].

For many years locomotives had been restricted to about 550 Rp and a wheel arrangement of 2-4-0 because of the dimensions of the existing turntables; this impediment was now removed with the introduction of the 4-4-0 locomotive, which was able, at a higher boiler output of initially 750 and later 1150 Rp, at twice the adhesive weight, and with improved running qualities, safely to haul heavy trains at speeds of up to 90 kmph (Fig. 4). The progressive introduction of the continuous, automatic Westinghouse brake, the improvement of the permanent way and of the safety measures had established further indispensable conditions for operations at increased speeds.

The elimination of the meal stops and the increase of the maximum speed to 80 and, exceptionally, to 90 kmph, permitted an appreciably higher speed level for the express trains. Thus train D31/32 Berlin–Hildesheim–Cologne reached about 64.3 kmph in 1892, and train D51/52 Berlin–Nordhausen–Frankfurt (Main) about 60.5 kmph in the following year. The new train became a great success, and seat reservations at DM 1–2.00 had to be introduced to avoid overcrowding of these popular trains.

The state railways did their best to introduce the new express train (Fig. 5) on all important long-distance lines and to cut down the journey times from one timetable period to the next; the fast trains on the less important lines were gradually reclassified as Eilzug (through trains) without supplementary tickets. There was a notable further increase in the weight of express trains through the addition of dining cars, through-coaches to day trains; and sleeping cars to night trains. It was nevertheless possible with the introduction of 4-6-0 or 4-6-2 locomotives to achieve a considerable increase in the service speeds up to 1914; top service speeds of 86.8 kmph were recorded on the Berlin–Halle run and of 88.8 kmph on the Berlin–Hamburg non-stop run. It

Table 1: Service speeds in kmph of express trains in 1900 and 1914 according to Meyer's Encyclopaedia 1904, "Die Reichsbahn" 1932 and according to timetables

	1900	1914
Berlin–Hamburg	79,5[1])	88,8
Berlin–Hanover–Cologne	65,9	71,0
Stendal–Hanover (without stopping)	78,9	82,1
Hamburg–Cologne	61,9[2])	64,8
Munich–Berlin	63,6	77,6[3])

[1]) stopping at Wittenberge
[2]) 1904
[3]) stopping at Nuremberg and Halle

Fig. 4: The brand-new Class S2 express locomotive heading one of the first express trains, photographed at Potsdam on 19th November 1893

should be borne in mind that only since 1902, with the introduction of new Operating Rules for the Main Railways of Germany, the maximum speed permitted for passenger trains with continuous brake, falling gradients of less than 5% (1:200) and radii above 900 m was generally 100 kmph. This was increased to 105 kmph a few years later.

Some express trains featured very long non-stop runs of the locomotives with few intermediate stops, such as Train D 79/80 Berlin–Munich, which stopped in Halle and Nuremberg only. This heavy express train nevertheless reached 77.6 kmph northbound, and anticipated the one feature of the long-distance express of the Twenties, ie the long non-stop runs. But before this next step in the development of high-speed transport is dealt with, it is necessary to describe the first series of high-speed records achieved shortly after the turn of the century.

High-speed records after the turn of the century

At the turn of the century international comparisons were a popular subject of the press and of public discussion. National pride which was flourishing at that time, made it particularly difficult to accept success by other nations, whether it involved size, comfort, and speed of ocean liners or the fastest trains. On rail the British and the French led the field.

In 1900 the British covered the 635 km between London and Edinburgh at a service speed of 83.2 kmph with three intermediate stops totalling 16 minutes, and on the 298 km non-stop run from Paris to Calais the French reached 85.1 kmph, and covered the 154 km from Paris to St. Quentin at 90.5 kmph. Their top trains were thus faster than the German ones; this did not, however, necessarily apply to the adjusted average of all trains. The magic figure of 100 kmph service speed, appeared within immediate reach.

Not one, but two initiatives became effective in Germany. The first was by the "Study Group for Electric High-speed Railways" founded before the turn of the century. With extraordinary enthusiasm electrical engineers had developed the invention of the electric railway from the air final, purely demonstration model at the 1879 Berlin Industrial Exhibition, through the first electric tramway at Lichterfelde in 1881, to a standard gauge railway. This was followed by an unexpected success, which must have looked like a challenge to the designers of steam locomotives. As early as 1901 a Siemens and Halske electric four-

Fig. 5: The express train, everybody's transport medium before the age of the motorcar; a holiday special leaving Berlin Stettiner Station for the Baltic coast

axle experimental locomotive achieved a world record of 162.5 kmph on 15 kv three-phase current on the almost straight 23.3 km military railway between Marienfelde and Zossen [6]. This was a break through of the magic barrier of 100 kmph, ie 160.9 kmph. And that was not all. Even higher speeds were aimed at with two power units with two three-axle bogies each.

Even at 160 kmph it became obvious that the rails were not up to the job of taking such high-speed traffic. One bogie was briefly lifted off the track, it bounced back and splayed the rails, a warning to all future experimental engineers to conceive vehicle

Fig. 6: The AEG three-phase railcar achieved a world speed record of 210.2 kmph on 27th October 1903

and track as an integrated whole, as "one machine", as old George Stephenson had already realized.

The speed trials could only be continued after a heavy permanent way had been installed; the success was striking: on 6th October 1903 the Siemens and Halske-equipped power unit reached a top speed of 202.7 kmph, and 206.8 kmph on 23rd October. A few days later, on 27th October, the AEG power unit (Fig. 6) advanced the record speed to 210 kmph. No less a sensation was caused by the record run of the Siemens electric locomotive hauling a train at 130 kmph.

These record runs had become possible only in experimental conditions. The contemporary three-phase technology used, with the complicated triple overhead wires, which were unsuitable for use in stations, was not yet compatible with practical operation. These record runs advanced practical experiments with single-phase a.c. in isolated operations in Berlin, Hamburg, Central Germany, and Bavaria. In 1911 the Prussian State Railways were able to demonstrate a train ride at 135 kmph to members of the Reichstag of the electrification of the Dessau-Bitterfeld line. The Germans thus held the speed records in all rail-borne categories [6].

With these record runs, however, electric traction had marked the level of future high-speed transport on rail and staked its claim to this forward-looking development, although this was appreciated by only few of the contemporary experts in the field. The leading positions in the state railways, in politics, and in the military were still occupied by men firmly rooted in the era of the steam locomotive.

Stimulated by the initiative of the electrical engineering industry and by the pride of the steam locomotice designers the Association of German Engineers launched a competition for a steam locomotive which would haul a 180 t train at 120 kmph and reach a speed of 150 kmph to make up any lost time.

Only after a second appeal designs were submitted which met the specifications of the competition. Only two 4-4-4 locomotives were built by Henschel & Sohn, Kassel, with 2200 mm dia. driving wheels, to the prizewinning design of Kuhn and Wittfeld [7]. Both locomotives of the Prussian Class S 9 had the aerodynamically designed footplate in front of the boiler (Fig. 7). Unfortunately, this Kuhn-Wittfeld locomotive reached only 137 kmph at the head of a 109 t train on the Lichterfelde–Zossen test track.

Hauling the same train, a mass production Prussian S 4 – a 4-4-4 locomotive with 1980 mm dia. driving wheels – attained 136 kmph. In spite of the fact that the S 9 was the most powerful express locomotive ever to run on Prussian State Railways; it was 1 m longer than the later Class 01. The S 9 did not take part in the subsequent trial runs of the Prussian State Railways between Spandau and Hanover, where the mass production locomotive S 7, a 4-4-2 saturated-steam 4-cylinder compound locomotive by Borries, reached 143 kmph at the head of a 156 t train [7].

These speed trials acted as a stimulus on the other state railways. As early as 1902 the Maffei locomotive works in Munich had delivered a 4-4-2 saturated-steam 4-cylinder compound locomotive to Bavarian State Railways which on a trial run was said to have reached 144 kmph. This experience enabled the Maffei designer Hammel to meet the order by Bavarian State Railways for a 4-4-4 superheated-steam 4-cylinder compound locomotive within only 4 months design and construction. This famous S2/6, like the S 9, had 2200 mm dia. driving wheels (Fig. 8).

Fig. 7: The Kuhn-Wittfeld Class S 9 high-speed locomotive of 1904 of the Prussian State Railway on a trial run reached 137 kmph heading a light train

Fig. 8: The fastest locomotive of the state railways, the Bavarian S2/6, record speed 154.5 kmph, at the Nuremberg depot 1984

On 2nd July 1907 this locomotive, with v. Frauendorfer, the Bavarian Minister of Transport, on the footplate, and hauling a four-coach train of about 150 t weight established a record of 154.5 kmph.

But the 2nd July 1907 not only represented a speed record on the road to the high-speed transport of today and of the future; it is also the day on which the performance of the express locomotives overtook the speed potential of the mainline railways. For especially in the German upland sections the main lines, most of them built between 1840 and 1880, have many curves with minimal radius of 2,000 ft (about 600 m), which in practice limits the speed to 100/110 kmph on 24 sections between Hanover and Wuerzburg alone.

In view of this, and in the absence of competition from airliner and motorcar, it was conceivable that the development of the steam locomotive was targeted on the heavier express train and on 110 or 120 kmph maximum speed. The solution was found with the heavy Pacifics, such as the Maffei 4-6-2 superheated steam 4-cylinder compound locomotive, the famous S3/6, at initially 1770 Rp or the Prussian 4-6-0 superheated steam 4-cylinder Class S 10 compound locomotive.

Soon afterwards the Great War put an end to the Golden Age of the state-railways, which had offered the German travelling public cheap transport in up-to-date express trains on long-distance lines across the length and breadth of the country.

August Scherl's high speed rail system

In May 1909, 6 months after Henry Ford launched his famous Model T and shortly before Louis Bleriot crossed the Channel in his plane, an unusual book was published by the August Scherl publishing company in Berlin, entitled "A New High-speed Railway System – Suggestions for the Improvement of Passenger Transport". It offered the first comprehensive concept of a network of high-speed railways in Germany. The author was August Scherl, the publisher himself.

Although August Scherl praised the advantages of the German state-railways in comparison with the foreign ones, he soberly analysed their shortcomings at the time: in the Ruhr district the provision of freight wagons did not meet the demand, many main lines were overloaded, and the passenger train timetable suffered because it also had to accommodate freight trains. In Scherl's view the express trains were too slow compared with those abroad and too infrequent. Scherl expected a further increase in the size of the cities as well as a superproportional increase of transport demands and came to the conclusion that within 12 years the state-railways would no longer be able cope with the volume of transport [8].

For passenger transport Scherl postulated a high-speed railway at 200 kmph, segregated from freight transport (Fig. 9), and disqualified the steam locomotive for this speed for plausible reasons, quoting Privy Councillor Borries: "At an average speed of 100 kmph the steam locomotive has reached the

Fig. 9: This 1909 study of Scherl's high-speed railway network anticipates the new lines of the German Federal Railway

limit of its economicals performance. Anything beyond is ostentation". Scherl referred to the Marienfelde speed trials and demanded electric traction with multiple units and all-wheel drive.

In considering the track, Scherl quite rightly recognized the considerable shortcomings of the main lines with radii of curvature that were often too small for high-speed operation, although his pessimistic assessment of a possible further development of the rail has now been outstripped by the introduction of the continuous welded rail and the much improved lining of the track. His negative prognosis induced him to suggest an elevated monorail railway gyroscopic stabilization. It was above all the unconventional nature of the system which prevented the adoption of the project; there were also other, plans for monorail systems advanced at the time, none of which, however, could be taken seriously.

Scherl demanded 5 things of the new passenger transport system: high-speed, safety, comfort, cost effectiveness, and continuity (frequency of service and connections). He designed an wide meshed multi-centric high-speed network, to be supplemented by a 120–150 kmph feeder network and a 30–60 kmph tertiary one. With this vision he obviously left reality behind. This also applies to his plans for the high-speed railway to be carried on steel trestles high above roof level (Fig. 10) to penetrate into city centres.

But a multitude of basic suggestions were practicable and stimulated current of designs, such as the complete segregation of the track from other transport arteries, especially from the roads, and track alignment in wide curves – the contemporary perspective drawing of train, track, and landscape already anticipates our future high-speed transport concept. We have the principle of the compact multiple unit conveying the overall impression already heralding the future high-speed multiple units. And when we consider the interior of the vehicle (Fig. 11), Scherl had already suggested those features of "travelling comfort" which our generation intends to introduce in the IC Express.

The design of a 4 m wide vehicle gange provided for compartments with 4 wide seats on either side of a centre aisle. Even more remarkable are a smoking saloon, hairdressing and writing saloons, dining room and a special cold buffet, as well as a central

Fig. 10: August Scherl's idea of a central station in a large city

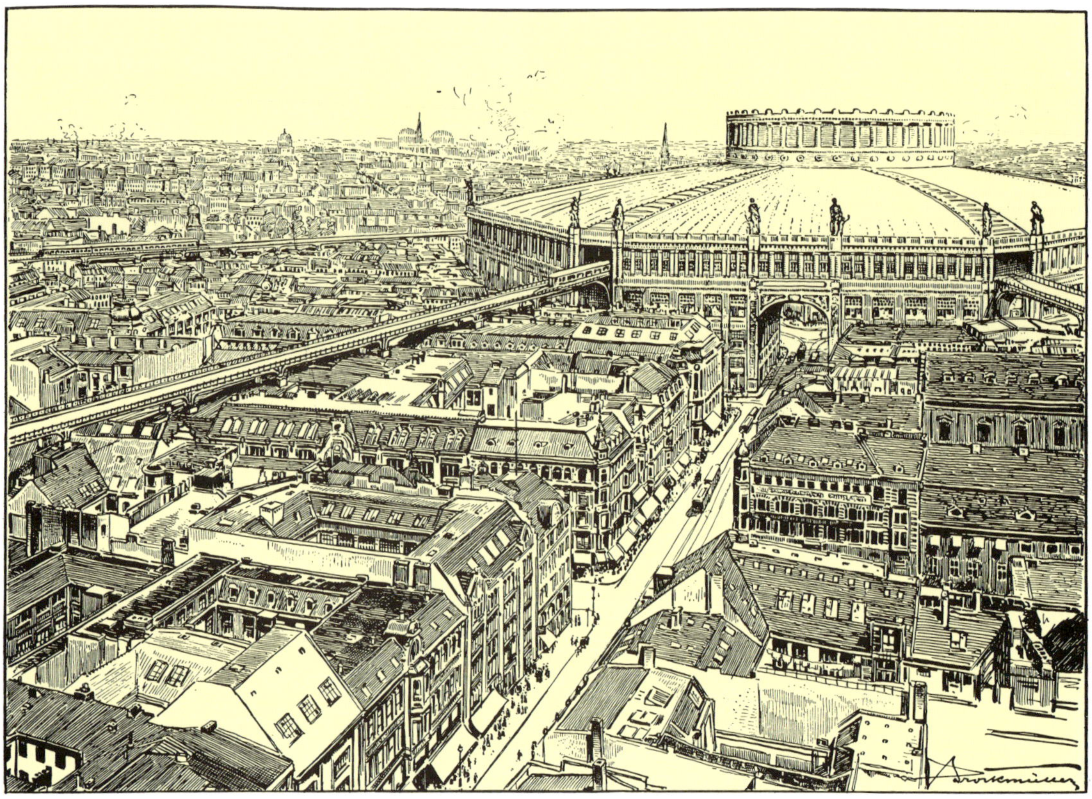

Fig. 11: View, in cross-section, and ground plan of the 3-car high-speed multiple unit of the August Scherl project 1909

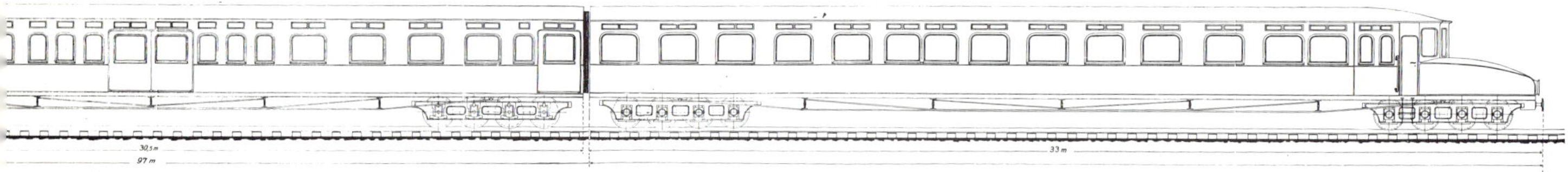

Fig. 1: Side view

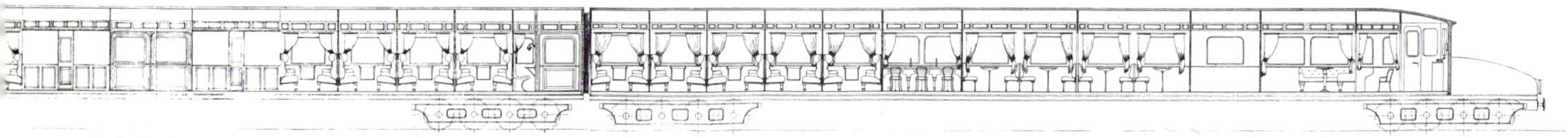

Fig. 2: Sectional side view

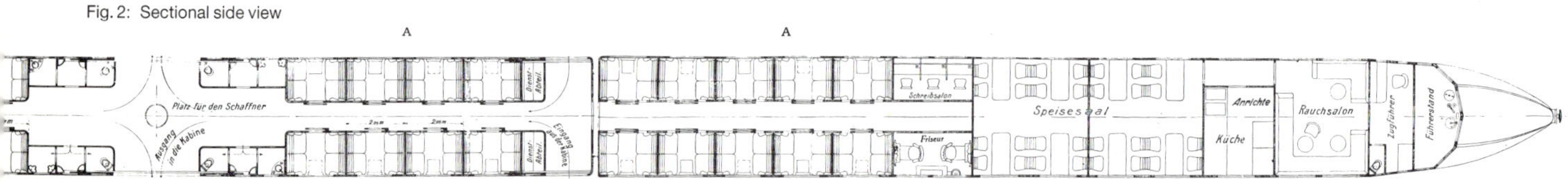

Fig. 3: Sectional plan view

Fig. 12: Club compartment in Scherl's high-speed railway, a contribution to the idea of comfort

entrance imitating the foyer of a hotel. Perspective drawings clearly demonstrate the aim of offering enhanced comfort to all passengers (Fig. 12).

Scherl wanted a fixed cycle timetable, suggesting a model of a 30-minute timing between trains, for instance for the Berlin–Hamburg line. He already realized that high speeds make running passenger trains at night redundant and planned to use the high-speed lines at night for express freight trains – another highly topical suggestion today.

All things considered, August Scherl was far ahead of his time. Many of his suggestions were partially implemented in the high-speed multiple units during the Thirties. But the nucleus of his project, the idea of a high-speed rail system, has until now been denied practical implementation, last, but not least because the Great War interrupted the continuity of peaceful development and civilization in Europe.

Long-distance express trains; the newly established Reichsbahn after the Great War

Four years of war accomplished what no passenger and no railwayman could have imagined: the once magnificient German railways system was decimated because of excessive strain in the service of the military and because of the lack of maintenance and renewal. Added to this was the handing over of the best locomotives and coaches to the Allies. After 1918 the maximum speed on German railways dropped to 75 kmph. It took until 1929 to regain the speed level of 1914. To provide a performance adequate for commercial transport during this decade of misery, of coal shortages and of hyperinflation, the "Deutsche Reichsbahn" established a new category of train, the "Fern-D-Zug" (FD-Zug, long distance train).

Two measures made it possible to improve the drastically reduced service speeds of the heavy express trains with the long-distance expresses: firstly, the planning of long non-stop runs, with intermediate stops in important conurbations only; and secondly, the reduction of the train load to three or four 1st and 2nd Class express coaches and a dining car, ie to 250–300 t. The trains were pulled by the powerful former Prussian Class S 10 and the Bavarian S 3/6 express locomotives.

The idea of a high-speed network and of same-day return trains which enabled businessmen to travel to Berlin from all parts of Germany in the morning and to return home in the afternoon, was new and indicated the direction for high-speed transport in Germany over the next 50 years.

A start was made on 1st July 1923 with Train FD 23/24 Altona Central station–Berlin Lehrter Station at 76.8 kmph service speed (from Hamburg Central station) and with Train Fd 79/80 Munich–Nuremberg–Halle–Berlin Anhalter Station at only 64 kmph, because of the mountainous route. The radial network was later extended to Berlin from Frankfurt (Main), Stuttgart, Cologne, and Beuthen [10].

Together with the entry into service of the new standard express locomotives, the fastest long-distance trains between Berlin and Hamburg regained the 1914 speed level of 88.8 kmph in 1929 after fifteen years; although admittedly, the long-distance expresses constituted hardly more than 5% of all express and through trains.

Franz Kruckenberg and his propeller-driven railcar

Several factors gave impetus to the development of high-speed transport during the Twenties and Thirties. There was, to begin with, competition from the aeroplane and the motorcar (Fig. 13). Added to this, however, were important technological innovation potentials: powerful, thermally favourable diesel engines, electric independent axle drives, hydraulic transmission, lightweight construction, (already far advanced in the airship and aircraft design), and finally, the developements in knowledge of aerodynamics for vehicle construction.

Fig. 13: Air travel casts its shadow, although competition with the railway will be taken up not by the airship, but by the aeroplane

In addition there were some far-seeing, and imaginative engineers, able to think boldly and in terms of new transport systems such as, above all, Franz Kruckenberg and his fellow pioneers, but also the management of the German Railway headed by Dr. Dorpmueller were responsible for much progress.

The aim of the Propeller-Driven Railcar Company Ltd (Flugbahngesellschaft mbH), entered in the Heidelberg company register on 5th August 1928, was the development of an elevated high-speed transport system. The managers were Franz Kruckenberg, the guiding spirit, and the capable Kurt Stedefeld, both engineers. The project of the GVT high-speed suspension railway developed by the Society for Transport Technology (Gesellschaft fuer Verkehrstechnik) under Kruckenberg and Stedefeld originated in 1923.

Kruckenberg described his concept of a high-speed railway as follows [9]:

"My definition of high-speed railways applies to trains considerably faster than the existing ones, such as the Siemens and Halske and the AEG high-speed units tested on the Marienfelde–Zossen line near Berlin more than 25 years ago, which reached more than 200 kmph. These efforts had to be abandoned because there were no solutions to the problems arising from high speed on rails.

At that time levels of scientific knowledge and of technology was still not sufficiently advanced. Today we know that lightweight construction, aerodynamic design, continuous-welded, perfectly smooth rails, and the highly developed technology of automatic signalling had to be established before a 200 kmph high-speed railway of technological feasibility and economic viability could be designed."

Because the prevailing financial and economic climate in Germany, with no government support for research and development, prevented the funding of a demonstration line for the propeller-driven suspension high-speed railway, Kruckenberg adopted the wheel-on-rail system in 1928 and teamed up with the German Experimental Institute of Aeronautics (DVL), which had already conducted airscrew experiments with a propeller-driven railway carriage, in 1927.

After further trials with these simple carriages, Kruckenberg built the "GVT propeller power unit", which became popular as "rail Zeppelin". On 23rd September 1930 the first high-speed trials were conducted on the unfinished 8 km Hanover–Celle line. The extremely lightweight (only 18.58 t), 25.4 m long, streamlined power car reached a maximum speed of 182 kmph.

The German State Railway had supported this development with permanent-way work and with provision of workshop facilities at Hanover–Leinhausen, as it was interested in the further developments of fast trains and as "the world's largest technological enterprise it will promote all technological progress" [10, 1930, pp. 1139].

After the first attempt at high-speed reflections by the German State Railway on future applications were published in the journal "Reichsbahn": "Although the propeller-driven power car is only undergoing its first trials, conclusions are nevertheless possible. From the safe, steady, and fast rides that have so far been observed on a straight track, it can be foreseen that a particularly fast light-weight vehicle can be developed on this basis. Its use may be feasible in stead of the steam locomotive, where fast and frequent services with short train sets are required to satisfy transport needs over medium and very long distances. Such fast railcars could be readily deployed for instance on the Berlin–Hamburg line, on a fixed-cycle timetable to provide a reasonable frequency of service to cope with the high-speed passenger volume between these two conurbations. Evidence that the German State Railway has been entertaining such ideas for some time is, shown for instance, in the high-speed service on the electrified Halle–Leipzig line with electric (multiple units) introduced in 1928..."

But the German State Railway also pointed at the Rate of development of the existing main lines for speeds of about 100 kmph, and later up to 120 kmph, and at the timetabling problem of accommodating 200 kmph lightweight multiple units between different categories of train on busy main lines. In 1931, the "Reichsbahn" [10, 1931, p. 637] wrote: "Fitting a high-speed railcar service into the timetable at maximum speed on a busy line such as Berlin–Hamburg creates serious difficulties. The simplest solution, building special tracks for the propeller-driven railcar, would incur extraordinarily high construction costs and is out of questions in our present financial situation."

Early in the morning of 21st June 1931, the propeller-driven railcar (Fig. 14) achieved a record run which far surpassed all existing speed records. From 03.27 to 05.05 hrs the railcar raced along the 257 km line from Hamburg-Bergedorf to Berlin Lehrter Station in only 98 minutes. A top speed of 230 kmph was reached between Karstadt and Wittenberge: only here was the engine power of 600 hp fully used. The low fuel consumption of an average of 70 litres/100 km was one advantage of the lightweight construction and of the internal combustion engine, over steam-hauled heavy long-distance trains [11].

Fig. 14: On 21st June 1931 the "rail Zeppelin" reached a speed of 230.2 kmph between the towns of Ludwigslust and Wittenberge. This illustration shows it at Spandau after the record run.

There was only one news item for Berlin and for the whole of Germany on 21st June 1931, the record run of the "rail Zeppelin!" And the German State Railway was already about to take the offensive, albeit in pursuit of a line of development which promised much better compatibility with the existing technological

65 kmph: Ancient Motor railcar (1924)

80 kmph: Modern Light railcar (1931)

100 kmph: Modern Express railcar (1932)

160 kmph: Streamlined High-speed railcar (1932)

Fig. 15: Developments in the high-speed multiple unit of the DR

and operational system, ie the diesel-electric drive, which industry had also offered to Kruckenberg.

High-speed multiple units – the first high-speed transport system on German railways

As far back as the Twenties, the "German State Railway" (DR) had made determined efforts to advance development of multiple units with internal combustion engines for use on branch line services. Due to its success, it was decided to develop a fast multiple unit on the basis of a 410 Rp Maybach diesel engine; construction started at Wismar in April 1930. In further developing of this multiple unit (Fig. 15) Dr Fuchs and Mr Breuer, high officials at the Berlin head office of the DR, designed a high-speed multiple unit of 41.9 m overall length designed for about 150 kmph maximum speed, (which would create fewer running problems than a 200 kmph vehicle), and with bogies, which promised a steadier ride than the single-axle Kruckenberg running gear. It also featured a diesel electric drive, well tested in conventional railway operation with 2 × 410 Rp Maybach diesel engines, with generators in the outer bogies and two electric motors in the centre Jacobs bogie.

In February 1931, the development and construction contract was signed with Wumag, Goerlitz, and in October 1932 the unit was already completed; on 19th December at 8 am, the high-speed multiple unit VT 877a/b started on its first test run on the Berlin–Hamburg race track. The presignalling distance was increased from 700 to 1,200 m, with the 1932 summer timetable and the fastest steam train of the DR accelerated to 91.1 kmph. After the town of Nauen, the multiple unit reached 150 kmph effortlessly, exceeded the legendary 100 mph, ie 160 kmph, and eventually attained a maximum speed of 165 kmph. On New Year's Eve 1932 the German morning papers reported about the official press run the day before, and that the British Swindon–Paddington express had to transfer the Blue Riband of the Rail. Derived

Fig. 16: First arrival of FDt 2 "The Flying Hamburger" at Hamburg Central Station on 15th May 1933

from the British "Flying Scotsman", the name "Flying Hamburger" quickly became popular [11].

At the start of the summer timetable on 15th May 1933, the DR introduced the first high-speed multiple unit into revenue-earning service. At 8.02 am, FDt 2 left Berlin Lehrter Station, and arrived at Hamburg Central station at 10.20 am after 2 hrs 18 min non-stop run over a 287 km-distance (Fig. 16). The maximum speed had been 160 kmph, the service speed 124.6 kmph [11]. In the event of technical breakdowns, in spite of a great effort, a Class 03 express steam locomotive with a replacement train set took 10 minutes longer.

Fig. 17: High-speed multiple unit at Frankfurt (Main) Central Station

Very soon the "Flying Hamburger" was so popular that seats were fully booked weeks in advance, in spite of the fact that the 98 2nd Class seats provided less comfort than modern express passenger coaches. The DR offered open saloons with facing seats on both sides of a centre aisle, with single seats on one side, and rows of three seats on the other side.

The remarkable success of the Berlin–Hamburg service and trail runs on other long-distance lines encouraged the DR management to build up a complete network of high-speed multiple-unit services on a day-return basis, centred mainly on the capital,

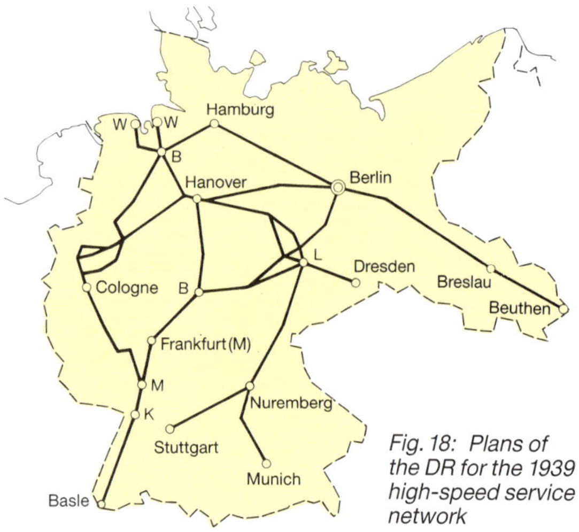

Fig. 18: Plans of the DR for the 1939 high-speed service network

Berlin, but also offering links such as Munich–Stuttgart and Hamburg–Cologne.

Based on the Flying Hamburger prototype (VT 877) of 41.9 m length, 13 high-speed multiple units (Schnelltriebwagen, SVT) of the Hamburg design were acquired in 1935; at a length of 44.2 m, they had 77 more comfortable seats in 2+1 arrangement in open saloons and a buffet. They were followed in 1935/36 by four 3-car SVTs of the Leipzig design with 2×600 Rp engines – two each in a diesel-electric and in a diesel-hydraulic version – at an overall length of 59.6 m and with 139 2nd and 3rd Class seats (Fig. 17).

14 3-car SVTs of the Cologne design with comfortable 2nd Class seats in compartments and a dining compartment entered service in 1938. In addition to these production units two SVTs of the Berlin design were added to the fleet; they consisted of a power unit with a slow-speed 1320 Rp MAN diesel engine and 3 coaches equipped similarly to the Cologne SVTs and with 155 seats.

In the 1939 summer timetable, with a total of 34 high-speed diesel multiple units and a few electric ones, the DR offered a high-speed multiple-unit network covering almost the whole of Germany, with the main junctions of Berlin, Hamburg, Hanover, Cologne, and Leipzig (Fig. 18). Of the 9271 km high-speed mul-

Table 2: Service speeds of the fastest train on DR-lines in 1914 and in 1927–1939

1	2	3	4	5	6	7	8	9	10	11	12	13	14	15	16	17	18	19	20	21	22
Consecutive No.	line			dis-tance	commercial speed													description of the fastest train			
	from	to	via	km	1914	1927	1928	1929	1931	1932	1933	1934	1935	1936	1937	1938	1939	1935	1937	1938	1939
1	Berlin Zoo	Hanover	Stendal	254,1	–	–	–	–	–	90,7	90,7	103,0	132,6	131,4	132,6	128,1	133,4	FDt 16	FDt 16	FDt 51	FDt 16
2	Hanover	Hamm (W)	Bielefeld	176,4	81,1	81,0	90,0	90,7	90,0	91,3	92,1	103,8	126,1	132,3	132,5	132,3	134,1	FDt 16	FDt 16	FDt 16	FDt 15
3	Berlin Anh Bf	Halle (Saale)	Wittenberg	161,7	86,8	83,1	86,8	86,8	90,0	90,7	93,3	105,5	106,6	104,3	101,1	101,1	122,7	FD 80	FD 80 FD 79	FD 80	FDt 33
4	Berlin L Bf	Hamburg	Wittenberge	286,8	88,8	82,0	86,1	88,8	88,8	96,1	124,6	124,6	124,6	125,6	125,6	122,9	125,6	FDt 2	FDt 2 FDt 1	FDt 1	FDt 24
5	Berlin Zoo	Cologne Hbf	Essen	578,9	71,0	72,9	76,2	77,5	78,7	80,8	83,9	90,5	116,9	118,1	118,1	113,9	118,1	FDt 16	FD 16	FDt 16	FDt 16
6	Berlin Zoo	Cologne Hbf	Gelsenkirchen	581,7	62,0	64,0	64,0	64,0	64,0	66,1	66,5	74,3	74,3	74,4	74,4	73,3	73,3	D 24	D 24	D 24	D 24
7	Berlin Anh Bf	Frankfurt (M)	Erfurt	538,9[1])	76,0	71,2	77,5	77,5	78,6	78,3	79,8	90,4	108,2	108,9	109,3	110,0	110,0	FDt 572	FDt 572	FDt 572	FDt 572/33
8	Berlin Anh Bf	Munich	Hof	652,5	76,0	59,6	62,0	62,0	63,7	64,4	66,4	66,6	69,4	69,0	70,2	70,2	70,2	D 23	D 23	D 23	D 23
9	Berlin Anh Bf	Munich	Nuremberg	674,2[2])	77,6	71,2	74,9	74,9	76,0	76,0	80,4	89,0	89,9	103,8	101,3	101,8	101,8	FD 79	FDt 551	FDt 551	FDt 551
10	Berlin Schl	Königsberg (Pr)	Marienburg	589,7	76,0	65,0	72,2	72,2	73,0	77,4	77,1	82,9	83,3	87,4	84,0	85,1	88,8	D 15	D 15	D 5	FD 5/6
11	Berlin Schl	Breslau	Sorau or Sagan	335,7	78,0	72,0	79,4	83,7	83,7	78,1	77,5	86,1	89,9	124,3	123,6	122,8	128,4	D 34	FDt 45	FDt 45	FDt 45
12	Cologne Hbf	Hamburg Hbf	Duisburg or Wuppertal	450,1	65,0	59,0	68,8	71,3	72,4	76,1	76,7	75,6	78,5	104,0	104,0	104,0	104,0	D 396	FDt 38	FDt 38	FDt 38
13	Frankfurt (M)	Hamburg	Kassel or Bebra	556,6	70,0	65,4	66,6	72,1	67,6	67,6	67,6	69,9	71,9	71,9	72,3	73,1	96,9	D 191	D 191	D 191/192	FDt 77
14	Frankfurt (M)	Munich	Ingolstadt or Augsburg	412,5	73,0	73,1	69,6	71,1	71,1	71,1	77,3	83,9	84,5	84,8	84,8	82,8	82,0	FD 263	FD 263	FD 263	FD 263
15	Frankfurt (M)	Basle DRB	Mannheim or Heidelberg	337,8	72,0	65,0	68,7	71,6	72,1	72,1	73,2	73,7	74,2	74,2	73,7	72,8	101,4	D 191	D 85	D 91	FDt 49
16	Munich	Stuttgart	Ulm	239,8	65,0	68,2	68,2	68,2	68,5	68,5	79,9	84,6	84,6	99,9	97,2	95,9	95,9	L 5	Dt 720	Dt 722	Dt 722
mean of consecutive Nos. 4–16				–	73,1	68,4	71,9	73,5	73,7	74,8	79,3	84,1	88,5	95,9	95,3	93,9	99,7	–	–	–	–

[1]) 550.0 (via Leipzig) [2]) 685.4 (via Leipzig)

tiple-unit network planned in 1934, the services not yet introduced were to Saarbruecken, Stettin, and Koenigsberg.

Table 2 [10] demonstrates the quantum jumps in speed achieved with the high-speed multiple units in that year compared with those of the fastest steam trains. Here one can clearly see the advantages of the mainly straight east-west alignments across low country in comparison to the winding north-south lines with long sections across the German uplands.

The DR used trial runs beyond the regular maximum speed of 160 kmph, to establish the absolute maximum speed of the high-speed multiple units. Thus the Flying Hamburger reached 170 kmph between Nordhausen and Berlin on 5th November 1933, and the Hamburg design the same speed between Bitterfeld and Wittenberg on 12th August 1935. On 17th February, a 3-car multiple unit of the Leipzig design achieved the record of all German Railway SVTs and the world record of all standard vehicles at 205 kmph between Hamburg and Berlin.

But the successful introduction of the first German high-speed transport system on rails must not solely be attributed to the application of progressive technologies in newly developed vehicles. The DR also improved the permanent way and the signalling system quite considerably. From 1927 onwards DM10m were spent per year, in spite of a difficult financial situation, on local line improvements (Fig. 19) to eliminate severe speed restrictions, and large-radius curved points were installed. Quite decisive, however, was the systematic strengthening of the permanent way for 20 t axle loads and the introduction of the 30 m rail. The increase in the presignalling

Fig. 19: The high-speed service network of DR was based on improvements of the whole system comprising vehicles, track, and operational equipment. Improved alignment of a station passage near Cologne

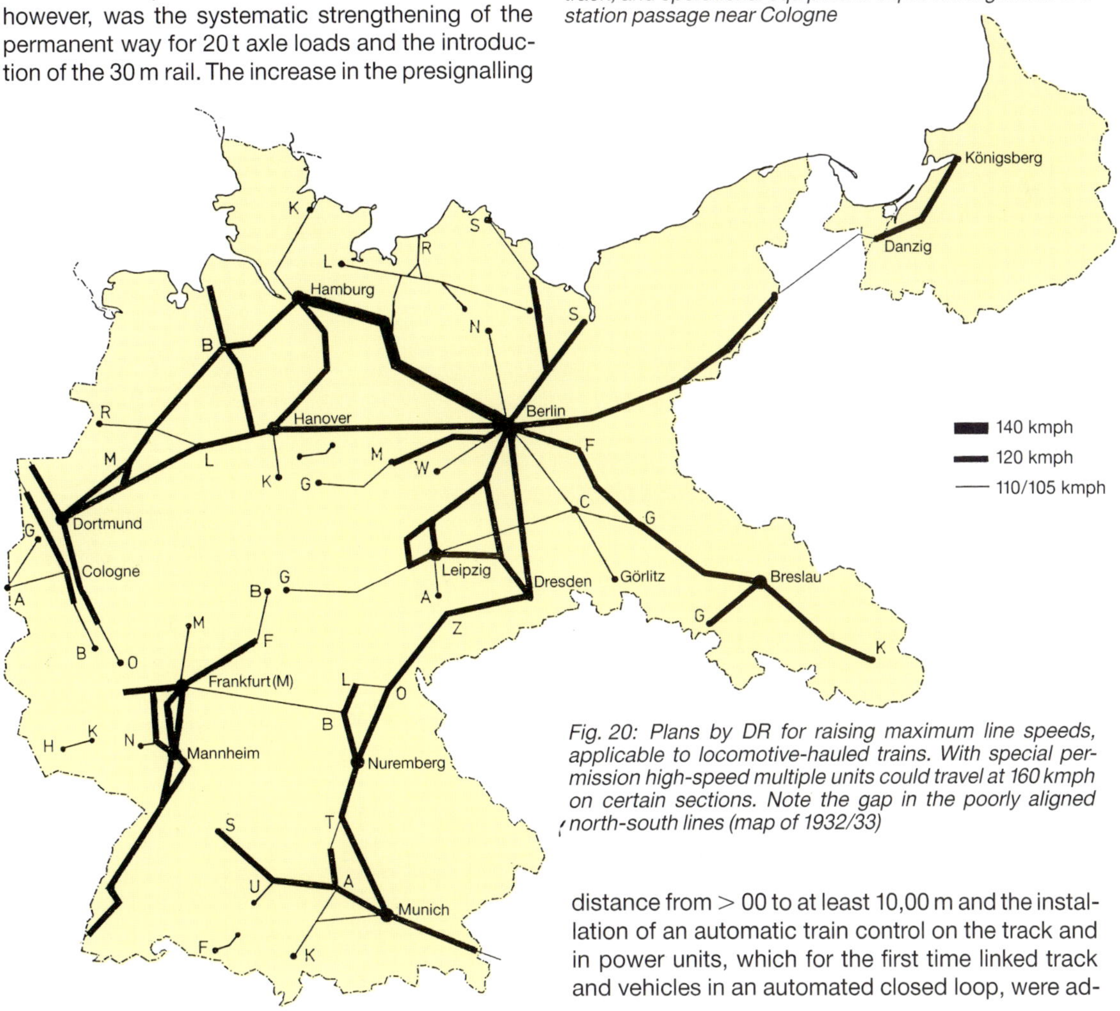

Fig. 20: Plans by DR for raising maximum line speeds, applicable to locomotive-hauled trains. With special permission high-speed multiple units could travel at 160 kmph on certain sections. Note the gap in the poorly aligned north-south lines (map of 1932/33)

distance from > 00 to at least 10,00 m and the installation of an automatic train control on the track and in power units, which for the first time linked track and vehicles in an automated closed loop, were additional improvements. All in all this was a major programme of modernization [10, 1934].

A map of the 1932/33 network is very revealing (Fig. 20). It shows the lines used by steam trains at 105 and 110 kmph maximum speed in 1933 and the planned increase in speed to 120 kmph in 1934 after the renewal of the permanent way and the increase in the presignalling distance, as well as the Berlin–Hamburg line, earmarked at the time for 140 kmph steam trains. The map shows the dominance of the east-west connections and the lower speed level the north-south ones across the barrier of the uplands. The high-speed multiple units occasionally ran at 160 kmph with special permission, for instance when they were fitted with a magnetic rail brake.

Technological competition on rails

The success of the diesel multiple units of the DR acted as a powerful stimulus on the chief designers of steam locomotives and of the electrical engineering industry. The centenary of the railway in Germany was thus characterized by the construction and trials of other spectacular high-speed vehicles, such as the three high-speed electric multiple units, designated elT 1900, built by the Esslingen Engineering Works and BBC and MAN and Siemens/AEG for 160 kmph maximum speed, the Henschel-Wegmann high-speed passenger train, and two Class 05 Borsig streamlined locomotives.

The elT 1900 high-speed electric multiple unit was one of the attractions at the Nuremberger Centenary Exhibition in 1935. A successful design, it was the first prototype of a high-speed electric multiple unit basically a predecessor of the ICE. With only three sets, it played the role of an outsider as a shuttle between Munich and Stuttgart during the Thirties, because electrification was only taking place at a leisurely pace at the time. To begin with, financial stringency had delayed the electrification plans of DR, so that only Nuremberg had been reached from the south. Then the Railway had to build motorways, and it may also have been objections raised by the army which prevented the implementation of a really comprehensive programme of electrification. However experience of bombing showed that overhead wires were often repaired more quickly than the network of tracks and points.

When Dr. Erich Metzeltin, the great locomotive expert of the firm of Hanomag, published his centenary volume "The Locomotive, too, is Celebrating" [16], he represented the steam locomotive as uttering the defiant words "as yet I would not dream of leaving the field to them. I remain what I am, the proud ruler of the rails".

Other designers of steam locomotives were inspired by the same sentiments. The locomotive manufacturers of Henschel & Son and the waggon builders Wegmann & Son, both of Kassel, suggested a particularly lightweight steam train for 160 kmph service speed for the DR. They envisaged a twin coach on three bogies, with 126 seats, and weighing 65 t, to be hauled by a streamlined 4-4-2 tank engine. This concept represented the type of steam train to compete with the 3-car high-speed multiple unit with the clear aim of reducing the heavy mass of the classic steam train through lightweight construction of the coaches and the lighter tank engine, and, to cut down air resistance, with streamlined fairing.

The DR realized that it would make little sense to rival the well-tried high-speed multiple unit directly with the steam locomotive, which is far less thermally efficient. It modified the concept towards a train with a higher seating capacity and 300 km range without

Fig. 21: The Class 05 001 high-speed locomotive of the German Railway, photographed in 1984. Its sister locomotive attained a record speed of 200.4 kmph at the head of a light train on 11th May 1936

Fig. 22: Franz Kruckenberger's 3-car high-speed multiple unit, the "Silberling" SVt 137 155, running speed 160 kmph

taking on water, ie towards a high-speed push-pull train for medium distances.

The "Henschel-Wegmann train", finally completed in 1935, comprised a streamlined 4-coach set, of only 125 t weight, with 192 2nd and 3rd Class seats, dining compartment and luggage space, and pulled by a 4-6-4 tank engine. This Class 61 was not only faired and designed for 175 kmph maximum speed but also had a footplate laid out specially for reverse running [10].

After display at the Nuremberg Centenary Exhibition and participation in the vehicle drive-past on 8th December 1935, the Henschel-Wegmann train was introduced to the press on a trial run from Berlin to Hamburg on 25th February 1936. It reached the maximum speed of 175 kmph effortlessly and a journey time of 2 hrs 32 min at 113 kmph service speed. This was convincing evidence of performance for the planned use of the train between Dresden and Berlin.

The press did the return journey in a special train of new express coaches hauled by the new Class 05 streamlined locomotive reaching 190 kmph on several occasions. Dr Wagner of the Railway Central Office and the Borsig engineers were the designers of this powerful high-speed locomotive. The aim of this development was to achieve, with these three-cylinder 4-6-4 locomotives with 2300 mm diameter driving wheels reaching 175 kmph maximum speed at the head of a 300 t train, considerably higher cost effectiveness than the high-speed multiple units with their limited seating capacity.

Both locomotives 05 001 (Fig. 21) and 05 002 were used to haul new long-distance morning- and evening expresses between Berlin and Hamburg in 1936, and in the Hamburg direction attained a regular service speed of 119.4 kmph.

On 9th and 11th May 1936, the DR demonstrated its high-speed vehicles to invited guests. The programme was extremely impressive. From Berlin to Stendal the guests travelled on the Henschel-Wegmann train, from Stendal to Hanover on a high-speed 3-car diesel multiple unit, from Hanover to Bremen on the Doble steam multiple unit, from Bremen to Hamburg on a 3-car diesel-hydraulic high-speed multiple unit, to return to Berlin behind a Class 05, dynamometer car, in the latest passenger coaches. This alone was an impressive demonstration of the technological developments in the high-speed transport, roots of which reached back into the era of the Weimar Republic. It also emphasized the ambitions and modern approach of the DR. But there was yet another climax to come: the official report in the "Reichsbahn" of 1936 states: the enthusiasm of all the passengers reached its climax on 11th May on the return trip from Hamburg to Berlin, when the message came from the dynamometer car that the train, consisting of passenger coaches of the latest design, with the streamlined Borsig locomotive at its head, had established a new speed record for steam trains at 201 kmph. To be precise, Senior Driver Langhans had coached his 05 002 to 200.4 kmph on a slight down gradient of 1:333, after the town of Zernitz near Nauen.

In 1938, the Kruckenberg 3-car high-speed multiple unit was finally completed by associated West German Waggon Works (Vereinigte Westdeutsche Waggonfabriken AG) (Fig. 22). Its special features: diesel-hydraulic drive by two 650 Rp engines housed in the projecting vehicle ends, systematic lightweight construction and up-to-date aerodynamic styling. To begin with the unit was dogged by misfortune and had problems with its axles.

However, on 23rd June 1939, a spectacular trial run took place: between Hamburg and Berlin, Kruckenberg's multiple unit reached a top speed of 215 kmph after the town of Ludwigslust. This was again a record for a vehicle designed for public service. Only 2 days later, this record run persuaded the DR to use the unit between Berlin Lehrter Station and Altona, to replace a failed SVT. On the return trip 200 kmph was reached between the town of Ludwigslust and Wittenberge when a jolt in the vehicle required emergency braking. An axle on a motor bogie had blocked. The unit was immobilized for many months.

Shortly before this AEG delivered the electric high-speed locomotive E 19. This introduced a development which eventually lead to the well-tried Class 103 high-speed locomotive via the Classes 110 and 111 of the German Federal Railway. Outwardly closely resembling the famous E18, the E19 at 5360 Rp = 3941 kW had a considerably higher continuous rating. It was designed for 225 kmph maximum speed, an indication that the DR and the design engineers expected the higher speed potential to reside in electric traction. The E19 was earmarked for the yet to be fully electrified Berlin–Munich line. The last E19 unit built by Henschel is said to have reached 225 kmph on trial runs [6].

Then the Second World War broke out. Among the countless, well-prepared orders of the first day of mobilization, was a ban on the movement of all diesel multiple units. The era of the German high-speed transport system thus came to an end overnight (Fig. 23).

The streamlined Class 01, 05, and 03 locomotives running on domestic coal continued in use (Fig 24). But they now hauled heavy expresses and leave trains from the various fronts for another six years until the German railway system was smashed under a hail of bombs.

Fig. 23: The German State Railway had every reason to publicize its safe, fast, and comfortable high-speed system abroad. At the bottom the Henschel streamlined locomotive of the Henschel-Wegmann train

The first postwar years, a repetition of railway history

After the Second World War, hundreds of thousands of officials and workers of the one-time DR set about the task of restoring the destroyed railway installations and of repairing rolling stock with primitive tools. During the first postwar years the railway once again passed, as it were, through its entire 150-year's history. To begin with, isolated Intercity connections were established in with jam-packed slow trains and coaches without lighting and without window panes, as in 1845. The first international through-trains and the first official domestic trains for privileged travellers across the zonal boundaries caused almost a sensation.

In August 1946, the official express train took 11 hrs 54 min from Frankfurt (Main) to Hamburg, recreating the situation of 80 years earlier. On 2nd October 1949, the ancestor of the high-speed multiple units, the "Flying Hamburger", entered service, under the operational designation SVT 04000, between Frankfurt (Main) and Basle as the high-speed multiple unit "Rhein-Main" FD 77/78. It reached 68.1 kmph service speed, approximately the speed reached at the turn of the century [12].

After 6 years of improvization, on 20th May 1951, the start of the summer timetable, the German Federal Railway offered a "network of the light-weight long-distance trains (Leichtschnellzüge)" to the expanding German business world for urgent commercial traffic.

Whereas in the pre-1939 high-speed transport network, the east-west traffic had dominated the situation, a network based mainly on north-south lines running, adapted to the multicentric structure of the new Federal Republic had now to be established.

From the point of rolling stock, the new high-speed transport system with a total of 15 connections was very heterogeneous. Similar to the long-distance expresses of the Twenties, the main category was long-distance trains hauled by Class 01, 03, 18 steam locomotives and consisting of two 1st and 2nd Class passenger coaches and one dining car. Added to these were some former high-speed multiple units which were gradually being relinquished by the Allies.

The average service speed of all high-quality passenger trains, ie the long-distance and international expresses, was 70.1 kmph in 1951, and after many improvements to bridges, permanent way, signalling installations, and rolling stock, at 79.7 kmph, almost reached the 80 kmph mark in 1958 [12].

The Dominance of diesel traction in long-distance high-speed transport from 1951 to 1971

Advancing northwards, electrification reached the Ruhr district in 1958 and the Hamburg region in 1965. Accordingly, a relatively rapid change in the type of rolling stock used for the high-speed long-distance expresses occured.

With the construction of the Class VT 08 diesel multiple unit the German Federal Railway (DB) resumed the multiple unit development of the pre-war era, but decided in favour of the diesel-hydraulic drive. This diesel multiple unit gradually took over the service of

the pre-war vehicles. Furthermore, the Class V200 diesel locomotive, first shown at the 1953 International Transport Exhibition, took over existing steam locomotive services in the long-distance train network (Fig. 25). In comparison with the steam locomotives, it offered greatly reduced pollution from fumes and smoke and interesting gains in journey times.

It was perhaps due to the limited line speeds of the north-south arterial lines through the uplands, that both the Classes V200 and VT08 were designed for only a 140 kmph maximum speed.

A few years laters, the planning of a network of fast and comfortable Trans-Europe Express trains created new development problems. Here, too, the specifications binding on all administrations, strangely, provided for more than 140 kmph maximum speed, although even then competition from air transport had become an already very powerful factor. The German contribution to this project was the TEE VT 11 multiple unit, later Class 601, the front of which recalled the design features of the Kruckenberg diesel-hydraulic high-speed multiple unit (Fig. 26).

By and large diesel power once again dominated the high-speed service of the German Federal Railway.

Fig. 24: In World War 2 streamlined steam-locomotives, too, hauled long distance express-trains whilest the high-speed multiple units remained in the workshop

Intercity trains at 200 kmph

At the beginning of the Sixties, the German Federal Railway (DB) adopted a new line of development for

Fig. 25: Diesel locomotive V200 hauling a light long-distance express, leaving Dammtor Station for Altona

Fig. 26: Diesel multiple unit VT 11, later 601, originally used as TEE, here as an Intercity train

high-speed transport, today's range of service developed. This line of development was characterized by electric traction, locomotive-hauled trains, comfortable coaches and 200 kmph maximum speed.

In 1962, the DB established the standards for comfortable long-distance travel during the next decades with a new type of coach for the "Rheingold Express". At the same time, the guidelines for 160 kmph maximum speed were published. In terms of maximum speed, the level of 1933 had at last been regained.

As the next step, the DB offered daily high-speed services at 200 kmph maximum speed between Munich and Augsburg, on the occasion of the International Transport Exhibition in Munich in 1965. It thereby demonstrated the efficiency of the railway and of the railway industry and set a new level of speed.

The following ten years were characterized worldwide by an optimism for the future which, originating in space technology, stimulated particularly transport technology. The future of high-speed transport on rails seemed to belong to unconventional support and control systems. On the other hand, the new Tokaido Line in Japan, operating at 210 kmph, and the record run of the French railway at 515 kmph in 1990, showed that quite considerable reserves of speed were still to be reached on the wheel-on-rail system.

These trends, which only can be described in passing here, stimulated two main developments:

▷ the consistent acceleration of high-speed transport on existing lines, and

▷ the start of systematic wheel-on-rail research running parallel to the development of magnetic levitation technology.

Fig. 27: Intercity train of the System IC 79, 1st and 2nd Class, headed by the E 103 high-speed locomotive in the valley of the River Main

The next milestone in the history of high-speed transport was the introduction on 26th september 1971, of the 1st Class Intercity system at regular 2 hourly intervals. With the new Class 103 high-speed locomotives and standard, comfortable, and air-conditioned passenger coaches, a standardized product was on offer for the first time since the war. It suffered, however, from two handicaps which were unacceptable in the long term. With its restriction to 1st Class the huge 2nd Class passenger potential, ie the average citizen, was ignored; furthermore, the permanent way permitted for the time being, a maximum speed of only 160 kmph. Nevertheless, at an average service speed of 104.5 kmph, the new IC system outpaced the TEE trains as the front runners.

Over the following years, the permanent way of the DB was effectively improved with the introduction of the heavy UIC 60 rail in combination with heavy and more narrowly spaced prestressed-concrete sleepers. Added to this was the start of programme of upgrading of the existing lines, which, in turn, was based on a master programme for higher speeds. The main aim was speeds of 200 kmph, or at least 160 kmph on extended sections of line.

It became possible at last to integrate efforts for improvement of the track, of electrification, and for the introduction of new rolling stock in conjunction with continuous automatic train control. Through the interaction between all the components of the railway system a jump in quality was achieved in high-speed transport with the introduction on 27th May 1979, of the IC 79 Train on a regular hourly interval basis (Fig. 27). For the first time not only businessmen and the well-to-do, but also the general public were able to enjoy 1st and 2nd Class high-speed travel. This was not only to their advantage, but also to the railway's advantage; because the extent of the IC service was considerably improved.

In view of this it was quite acceptable that the heavier trains reduced the service speed on the network from 105.3 kmph (1978) to 102.3 kmph (1979), because the savings in time for the 2nd Class passengers, who had hitherto been confined to the slower trains, was much more important.

The significance of the Germany Intercity network to regional planning is generally underrated: the premier trains of other railways may be faster, but there is no comparable system promoting a multicentric and sound regional planning structure nationwide, as effectively as the IC system, with its 36 regular and 11 temporary terminuses. This advantage must be preserved also in the ultra-high-speed system of the future.

In the 150th anniversary year of the railway in Germany, IC 85 introduced further improvements in high-speed transport, which largely exhaust the present-day potential of lines and of rolling stock. Through extension of the 200 kmph sections (Fig. 28) to a total of 440 km and tight schedules, it has been possible to raise the average service speed once again – to 108 kmph. The front runners are the IC 582, 680, and 682 trains in the Hanover–Hamburg section at 115 kmph service speed. But another aspect of IC 85 should perhaps be rated much higher: a new departure in passenger service: guards in charge of the train and conductors, who used to change trains every so often, have been replaced by IC chiefs and IC attendants who assist and serve the passenger. At the same time, the number of train crews has been increased by almost one third.

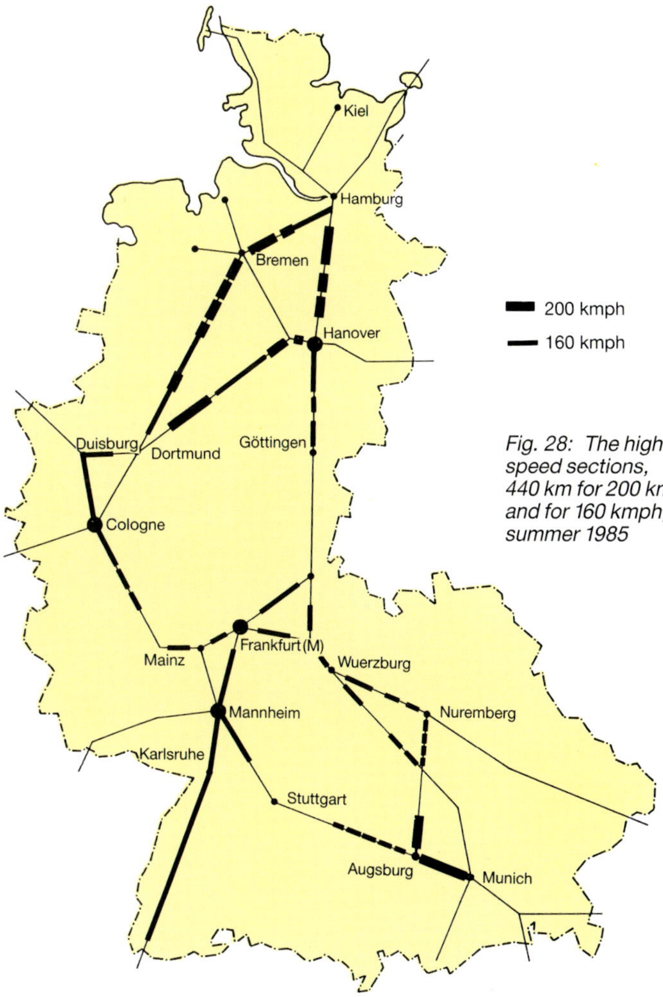

Fig. 28: The high-speed sections, 440 km for 200 kmph, and for 160 kmph; summer 1985

On the threshold of ultra-high-speed transport

Whilst the IC 85 concept registered an increase in traffic; planners, designers, and civil engineers were able to harvest the fruits of many years of technological development in the fields of vehicle technology and of the construction of new lines.

At a top speed of 265 kmph the Class 120 electrical locomotive had, on 7th October 1984, already established a speed record for locomotives with three-phase drive, and thereby emphasized the extremely positive future prospects for three-phase technology (Fig. 29). At the time, the ICE multiple unit, based on the same three-phase technology, had already reached the final assembly stage. It the Class 120 locomotive was conceived as an all-purpose traction unit for both passenger and freight transport, the superior, streamlined ICE was from the beginning the vehicle of the coming ultra-high-speed transport era. On 26th November 1985, before the anniversary year came to an end, the 5-car ICE reached a top

Fig. 29: Three-phase locomotive, Class 120

speed of 317 kmph on the race track of the Thirties, between the towns of Guetersloh and Neubeckum. However the potentialities of the existing lines were now exhausted.

Only the opening of the new line between Fulda and Wuerzburg, with its sweeping alignment and increased tension of the overhead wire for the trial run, permitted an even greater jump into the future: on 1st May 1988, Dr. Warnke, Federal Minister of Transport, Dr. Riesenhuber, Federal Minister of Research and Technology, Dr. Gohlke, Chairman of the Board of the German Federal Railway, and leading personalities from industry and from the Federal Railway, together with representives from the media and last, but not least, driver Romano Henning had their eyes glued to the speedometer with anticipation. Seemingly without effort the ICE spurted over 300 kmph. With a comfortable ride and low noise, only the speedometer and the catenary supports racing past provided indications of the speed. There was a burst of enthusiasm on a down gradient in the Muehlberg Tunnel 403 kmph were reached. It felt like flying across the valley of the River Main. The rising gradient in the Einmalberg-Tunnel slowed the ICE to 400 kmph, but in the Sinnberg-Tunnel the speedometer indicated the maximum speed of 406.9 kmph (Fig. 30).

The "Blue Riband of the Rail" changed lands: from the French TGV (380 kmph on 26th February 1981) to the ICE!

DB have now ordered 60 train sets for the ultra-high-speed service from 1991 onwards. Measured by its seating capacity the ICE corresponds to 6 diesel multiple units of the "Flying Hamburger" type. Never before in the history of the German railways had the speed level been raised by such a large jump, ie from 160/200 kmph to 250 kmph maximum speed. In 1991, 427 km of new lines, ICExpress trains on a regular-interval timetable, with great comfort; new forms of signalling technology and operational control will usher in a new epoch for the German railways, the era of ultra-high-speed transport.

Fig. 30: On 1st May 1988, the Intercity Experimental (ICE) "breached" the "sound barrier" of speed on rail (400 kmph) for the first time. At 11.12 am the speedometer in the train pointed at 406 kmph. Subsequent evaluation produced the precise value of 406.9 kmph

Dietmar Lübke

The ICE system

Consideration of, and alteration to the overall system

Within the framework of the "New Railway" strategy of the German Federal Railway (DB) a target model has developed: to stop the negative trend in market development away from long-distance rail passenger transport, to make possible a selective increase of commercial market share through expansion of marketability, whereby an improvement of the competitiveness of the railway in long-distance passenger transport can be achieved.

The opportunities for the DB in long-distance rail passenger transport are favourable. DB has realized that the current social climate offers possibilities it can exploit: more and more people are making use of increased services. The railway can take advantage of this situation by developing into a service enterprise and by accepting the challenge of competition, where it enjoys equal opportunities. To succeed in this endeavour, the railway must give increased consideration to its customers' wishes. It must be quicker and more flexible in its reaction to market forces. The existing innovation and development potential must be fully realized.

Unmistakable brand names represent the quickest route to success. The most important point to consider here is consistent emphasis in the market place on the strong advantages of the railway – punctuality, generous space, comfort, communication, and scenery. To increase their impact, the products must be adapted in good time to market requirements and to the competitive situation, through a product relaunch according to the product life-cycle idea or, if they can no longer survive in the market, replacement with new products.

The offer of services on the one hand, the breakthrough of the wheel-on-rail system into a higher speed range, the increasing complexity of railway operation in general and the interdependencies of the principal parameters on the other, make consideration of and attraction to the overall system essential. For future high-speed rail transport, this means the introduction of new rolling stock as well as innovations and improvements above all in network and tracks, accessibility, and service, as well as frequency of service and punctuality.

Fig. 1: Hanover–Wuerzburg New Line

Fig. 7: Remote-controlled train indicator boards show the passenger where to find his train

a rail journey. For this facility not only includes the reservation of a parking space in, or in the immediate neighbourhood of, the departure station (Fig. 6) but also the possibility of booking a hire car or taxi, at the destination station by telephone from the train. In addition to this, hire cars or taxis can be booked directly by computer at more than 500 ticket offices and at more than 1000 travel agencies.

Another example of services the improving accessibility is luggage transfer from station to airport under the name of "Rail and Fly". Luggage transfer to Frankfurt/Main Airport is accurrently possible from more than 60 stations.

Another aspect favouring travel by InterCityExpress, which should not be forgotten, is the station as the link between the ICE journey and access to the ICE system. The main features here are the functional facilities and the image presented by the station.

Long walking distances may have to be covered between arrival at the station and the seat on the ICE, particularly at large termini. Added to this is time spent on obtaining information and tickets, seeing to luggage, and on the purchase of articles necessary for the journey. All structural criteria for modification of a station must therefore have "short distances" as their main aim.

To reduce time and effort spent by the passenger, the installation of mechanical aids such as escalators, lifts, or luggage conveyor belts is desirable. In many large stations an amenity which had been neglected for decades has now been revived: porter service. This functions by booking by postcard or by telephone – from the moving train. The replacement of stairs by ramps benefits above all disabled passengers, and those with luggage on trolleys.

A further aid consists in providing as far as possible direct lines of sight between the passenger and his destination. Where the ICE is not directly visible, a lucid and unmistakable guidance system with pictograms, directions, and ICE-specific colour codes must come to the traveller's aid. On his way to the ICE, the passenger requires information about its whereabouts. Remote-controlled train indicator boards (Fig. 7) and train destination boards give valuable service here.

The ICE system also includes passenger service "from a single source". This must include informa-

Fig. 8: Up-to-date customer service at the Wuerzburg Central Station Travel Centre

tion, ticket sales, and seat reservation, possible hotel reservation at the destination and other services. The increasing number of travel centres equipped (Fig. 8) with the modern "KURS '90" computer-assisted facility offer such services. "KURS '90" is a versatile, customer-friendly travel, information, and sales system, which is connected with the railway computers.

One component of the future ICE system must also be a fare concept which is easily understood. Cashless payment by means of credit cards and through tickets for ICE/public short-haul passenger transport come to mind.

Another feature of the ICE System is a clean and relaxing travel ambience. The need to look after the individual passenger rather than control and surveillance must in future be the main task of the train crew (Fig. 9). This means above all information on board the train, assistance with luggage and with locating the booked seat, but also on-the-spot catering for the passenger, with cold and hot drinks as well as with snacks and light meals. In addition, this provides better seat utilization through seat allocation, fulfilling the passenger's wishes, more revenue through thorough ticket checks, reduction of deliberate damage caused by passengers, and increased punctuality through assistance with boarding and alighting of passengers and with train dispatch; this ensures that scheduled stopping times are complied with.

Service frequency and punctuality

First and foremost the passenger expects great punctuality, reliability, and an adequate frequency of service of the future ICE system.

The frequency of the regular-interval service must depend on the demand for transport, the volume of traffic, and the profitability of the operation. Market analyses have shown that the traveller in the up-market long-distance sector wants at least a one-hour regular-interval service. Traffic peaks must be accommodated with a shortening of the regular interval. Since the timetable change on 29th May 1988, with the opening of the new-line section be-

Fig. 9: Individual attention to the passenger on the train: head of a train crew of the German Federal Railway advising a passenger

Fig. 10: Computer-assisted train running control Stuttgart Central

tween Wuerzburg and Fulda, the IC trains of Line 4 have saved almost 30 minutes on the journey time. This has already made it possible to introduce a 30 min-interval service between Hamburg and Fulda between Lines 3 and 4. Through the existing system links it was also possible to pass on the gains in journey times to the rest of the network.

Further demands to be made on the future ICE system are an increased number of direct connections without change of trains, point-to-point connections without intermediate stops, and synchronized connections at the change-over stations and termini, with the changing of trains made easy across the same platform.

Of great importance to punctuality is the consistent use of radio control of trains for disposition and elimination of irregularities, on the one hand, and on the other, for integrated centralized traffic control and train control, which continuously feeds information about the location of the ICEs to the signalboxes and open-line installations and transmits it to the appropriate computer centre (Fig. 10).

Punctuality of the ICE system is further increased by the establishment of ICE reversing facilities, if possible in the immediate vicinity of the termini, with installations for cleaning, waste disposal, minor maintenance, and supplying the train with service material; and catering stores.

Punctuality of the ICE system can also be improved with the segregation of fast and slow trains through appropriate timetabling. Here, the location and design of the passing stations must be considered.

System-specific maintenance

Another component of a high-speed system is the maintenance of the vehicles and of the infrastructure.

The service periods of the series-produced ICEs is based on a limiting value of 70,000 to 80,000 km performance per period and an annual performance of 400,000 to 450,000 km. Considerations for vehicle maintenance in the repair shop can be based on a limiting value of 1.2 m kilometric performance for bogies and coach bodies, on a service life of about 30 years, and on the annual performance already mentioned.

Fig. 11: New motive power depot at Hamburg–Eidelstedt

Fig. 12: The new InterCityExpress train generation

Original intentions to adopt individual coach maintenance for the ICE trains were abandoned in favour of an integrated treatment of the complete multiple units. This cleared the way to an innovative concept which promises a considerable measure of success in rationalization. The building of the new motive power depot at Hamburg-Eidelstedt is concrete evidence of this concept (Fig. 11). The possibilities of the wide-band communication addition make any necessary maintenance of the ICE units considerably easier. The ICE motive power depot will learn of possible defects even whilst the unit is running and can thus prepare for any repairs before the unit even arrives.

On the permanent way maintenance sector, too, the reduction of cost is a principal aim. This involves the establishment and use of a system of rating numbers that serves as a criterion of safety, comfort, and the planning of maintenance work. An important step towards the essential objective assessment of the geometrical conditions of the track has now been taken with the introduction of analysers in self-propelled track inspection trains, which process the geometrical measurable variables and record them in the form of ratings for the assessment of the need for maintenance work. It is desirable to develop track assessment, hitherto based on geometrical data, so that actual stresses occuring between vehicles and track can be directly derived from these data as the most important safety criterion.

A new dimension of travelling

From 1991 onwards, after the opening of the first two new lines, Hanover–Wuerzburg and Mannheim–Stuttgart, the first high-speed connections in the Federal Republic of Germany at a scheduled maximum speed of 250 kmph on the basis of the ICE system will be established on the routes Hamburg–Hanover–Wuerzburg–Munich and Hamburg–Hanover–Frankfurt–Mannheim–Stuttgart–Munich. The 41 multiple units required for this service are already under delivery, and the first complete ICE trains have already started pre-trial operations.

Upon completion of new construction and of upgrading on the Karlsruhe–Basle line, a third ICE line, Hamburg – Hanover – Frankfurt – Mannheim – Basle, with a further 19 ICE trains, will be established.

Expectations of the success of the future ICE system are great indeed: through the considerable savings in journey times, the customer-friendly furnishing and layout of the vehicles, facilities and range of services; an enhancement of the image of the railway is to be achieved, which will gain new customers for this mode of transport. This means a considerable increase in the competitiveness of the railway, through lengthening the competitive radius to 400–600 km in comparison to other methods of transport.

In addition, the ICE system is designed for international traffic. Destination stations could, for instance, be Vienna, Berne, Zurich, and Geneva. The ICE fulfills all the conditions for such a service: it uses same current system and equipment of the power units with SBB pantographs, design of the steps for the international platform height of 550 mm above top of rail, and width and height of the multiple units adapted to the Gotthard-Tunnel profile.

Furthermore, the development of an ICE/M multi-system train has started and features the most advanced technologies; given unimpeded progress, prototypes will be ready for tests at the beginning of the Nineties and this will introduce yet another a new generation of European high-speed trains.

Gerhard Pintag and Josef Wiese

ICE – High-tech on rails

Product of 10 years of research and development

Since 1972 the Federal Ministry of Research and Technology has been promoting research and development projects in the sector of railway systems. By the end of 1989 it had spent about 500 Million DM on the wheel-on-rail technology sector. Industry and the German Federal Railway (DB) contributed considerable additional funds of their own, because the rate of government subsidy varies between 30 % and 100 % depending on the nature of the various projects.

A "Study on a high-performance high-speed rail system" reinforced other initiatives at the beginning of the promotion programme. One of its results was the realization that a high-performance high-speed railway makes sense within the distance – and speed range within which the railway is at present too slow, and the plane too uneconomical. The study recommended the establishment of efficient connections between centres of population at speeds from 300 to 400 kmph to close an existing speed gap in the overall transport system of the Federal Republic of Germany and of Central Europe.

For a lasting improvement of the basis for decision-making on long-term transport investment the Ministries of Research and Technology and of Transport are collaborating closely on the selective promotion of research and technological development of transport and traffic.

The following aspects occupy the foreground of this promotion:

▷ technological contributions to the establishment of efficient transport infractructures and services, improvement of the quality of transport facilities,

▷ lowering of the total cost of the transport services, thereby improvement of the economic situation of the transport operators,

▷ increase in reliability, safety, sensible energy and raw materials utilization combined with the greatest possible environmental compatibility of the transport systems,

▷ assurance of the long-term competitiveness of the public transport operators and the export industry,

▷ safeguarding jobs and availability of one of the most important instruments of the structural policies of regional planning, of housing, and of urban development.

Two guidelines are pursued in the railway sector, the improvement of the existing transport facilities for the full utilization of its performance reserves and the development of selected novel concepts, systems, and components, which in this context concentrates on the promotion of the maglev systems.

In the wheel-on-rail research programme the elements of the rail transport system have been systematically revised on the basis of up-to-date technological and scientific aids; the ICE, which in a maximum performance version will demonstrate the technological level reached, is now introduced as the comprehensive outcome in the vehicle sector of the system.

Execution of the wheel-on-rail research promotion programme

The technological basis of the existing railway system was established during periods when scientific knowledge and working methods available today were beyond the wildest dreams of planners and designers. But the railway system offered the advantage of being able to provide in the simplest technological a high transport capacity, which made industrialization and even the development of other transport systems possible. It was therefore necessary to explore, as a first step, the technological and economic limitations of the system to recognize the future performance aspects a high-efficiency railway can be identified with.

The programme was structured according to various sectors of technological disciplines:

▷ the sector of interaction between vehicle and track, in which in addition to the physical basic research of the support and guidance system, the analysis of the vehicle/track dynamics, and theoretical investigations of the advance calculation of practically optimized systems are carried out,

▷ the vehicle sector where components for high-speed vehicles are designed, novel drive and brake technologies and the associated control concepts developed, aerodynamic effects investigated and novel vehicle concepts established,

▷ the track sector, concerned with the design of components for high-speed tracks, the investigation of novel forms of permanent way, the analysis of the dynamics of track structures, and the optimum design of soil and foundation engineering,

▷ the sector of operating control technology, which deals with the analysis of operational dynamics, the

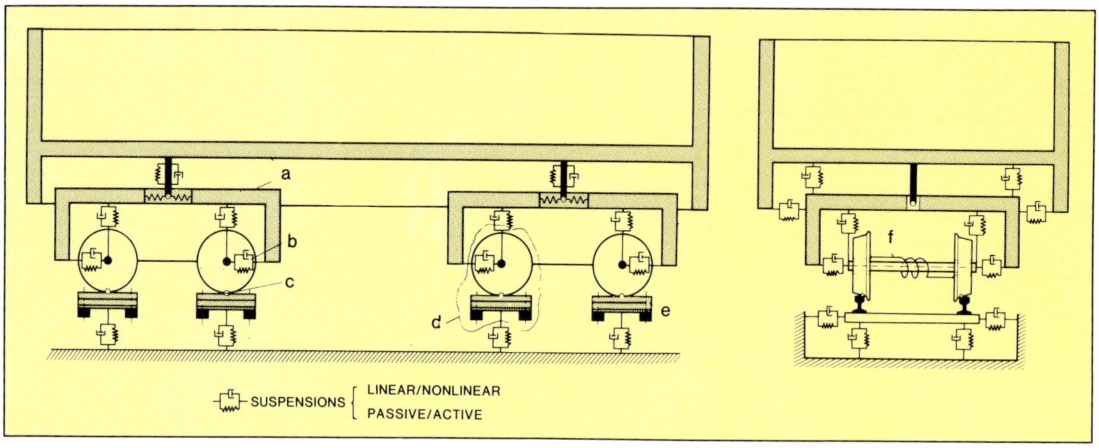

Fig. 1: Analytical model of a passenger coach

a = body rigid or flexible
b = coupling elements
c = non-linear contact mechanism
d = substructure
e = track panel element
f = elasticity of torsion of the wheelset axle

establishment of concepts of operating control, the design and layout of components for operating control systems, the synthesis and testing of structures for operating control systems, and questions of safety and reliability, which must be approached especially in the context of the use of new electronic technologies,

▷ the power technology sector, which covers the design and layout of power supply components, the analysis of the pantograph/contact wire dynamics, and the establishment of new concepts of power supply and technology,

▷ the sector of environmental questions, on which analyses of the mechanisms of noise, mechanical, electrical and aerodynamic effects are carried out, measures to minimize the environmental impact of vehicle-, track-, and power supply components worked out and human stress and its limits and the stress on the environment is investigated.

Interaction of vehicle and track

On this sector the central functional group of the railway, i.e. the support and guidance system, is investigated. It aims at making all the physical-technical effects amenable to a mathematical description, so that optimum structural solutions can be calculated in advance by means of appropriate assured mathematical methods.

In addition to the experimental exploration of physical effects that cannot yet be mathematically expressed, such as the reactions in the wheel contact area during rolling, the description of the concurrence of the mechanical elements, for instance in the running gear, is essential if the objective of the investigation is to be reached. An analysis of the structural principles leads to the design of a model which is expressed directly in mathematical equation systems, and in which real structures are imitated with a suitable combination of mathematically defined model elements. With correct representation the mathematical results of this equation system correspond to the behaviour of the real structure. Here not only the overall behaviour but, for the structural layout of the individual elements of the structure, the behaviour at all interfaces to be found in the real structure is of interest. For the verification of the correctness of the model confirmatory experiments are necessary; some of these can be carried out on the line (overall information), but others must consist of selective laboratory tests in defined reproducible conditions (element information). Suitable test rigs have been designed for all these tasks within the research programme.

Fig. 1 illustrates an example of the model of a passenger coach as used in the programme systems established in the research programme. Such models serve for the simulation, assessment and dynamic layout of passenger coaches, which are regarded as multi-body systems here.

In Fig. 2 an example is shown of the calculation of the effect of individual structural elements, such as the active hunting control in the bogie. The point of intersection between curve 1 and the "imaginary part of the Eigen-value" axis indicates the limit of a stable wheelset running. In this example this is reached at about 280 kmph for the passive hunting control.

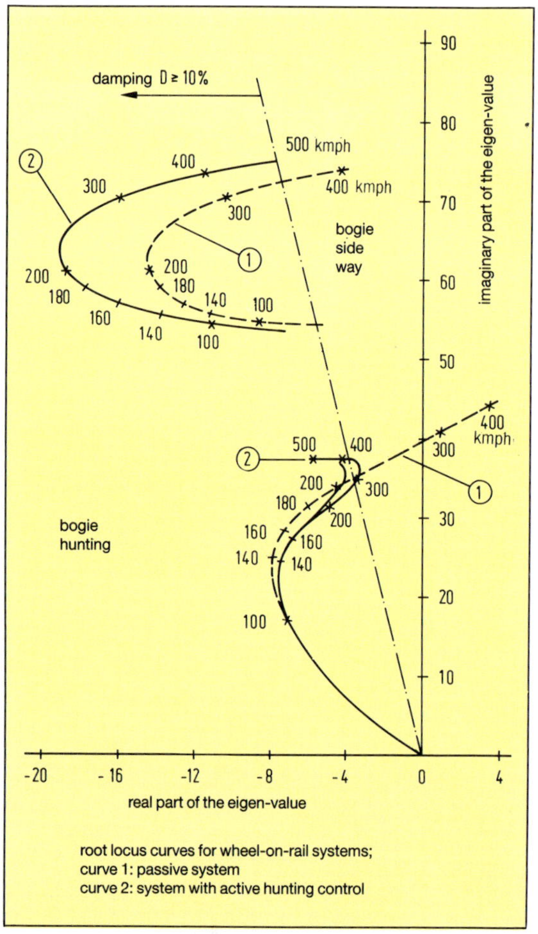

Fig. 2: Effect of structural elements on wheelset running

When active hunting control, represented by curve 2, is applied, the limit of the investigated speed range was never reached up to 500 kmph. This behaviour was confirmed on the roller test rig, of which more later.

One of the programme structures used in the research programme for the non-linear simulation of a wheel-on-rail vehicle is illustrated in Fig. 3.

In addition to many other actuating variables the effect on the whole vehicle of cross winds and gusts, affecting the flexible coach body in the centre of the picture, can be calculated here.

The theoretical information gained has been summarized in an overall computer program, whose basic structure and progress is illustrated in Fig. 4.

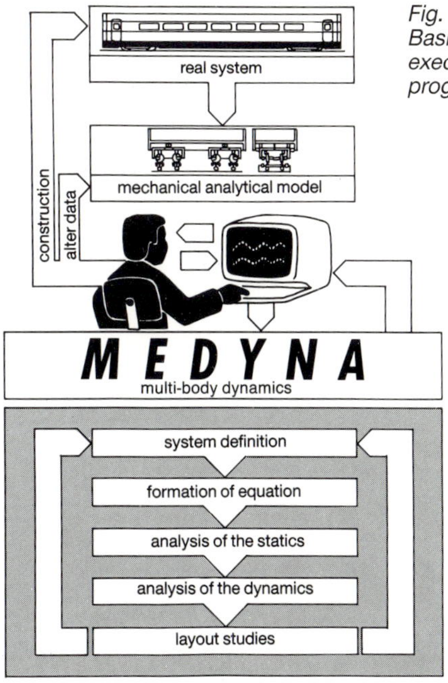

Fig. 4: Basic structure and execution of the MEDYNA programme

The roller test rig

The roller test rig built at München-Freimann is an important experimental instrument, on which fullscale experiments can be carried out in defined, reproducible conditions. Construction started in 1974, two-axle operation in 1977. Since 1980 it has been available for four-axle operation also for the testing of complete vehicles. The installation is shown in Fig. 5. The roller test rig offers the facility to investigate the runnit behaviour of vehicles. It also permits the study of the influence of individual components such as springs, shock absorbers, and mass distribution on running behaviour. For this purpose various running states with variable geometrical positions in space of the wheel contact areas can be simulated with different utilization of the coefficient of friction with the adjustable rollers. In addition, a real track with its faults spread in time can be simulated. Among the geometrical parameters track gauge play, rail cant, transverse and vertical state of the track bed, the rail section, and random combinations of these values can be set and varied in time. Rollers and wheelsets can be driven to generate variable driving as well as brak-

Fig. 3: Example of the structure of a simulation programme

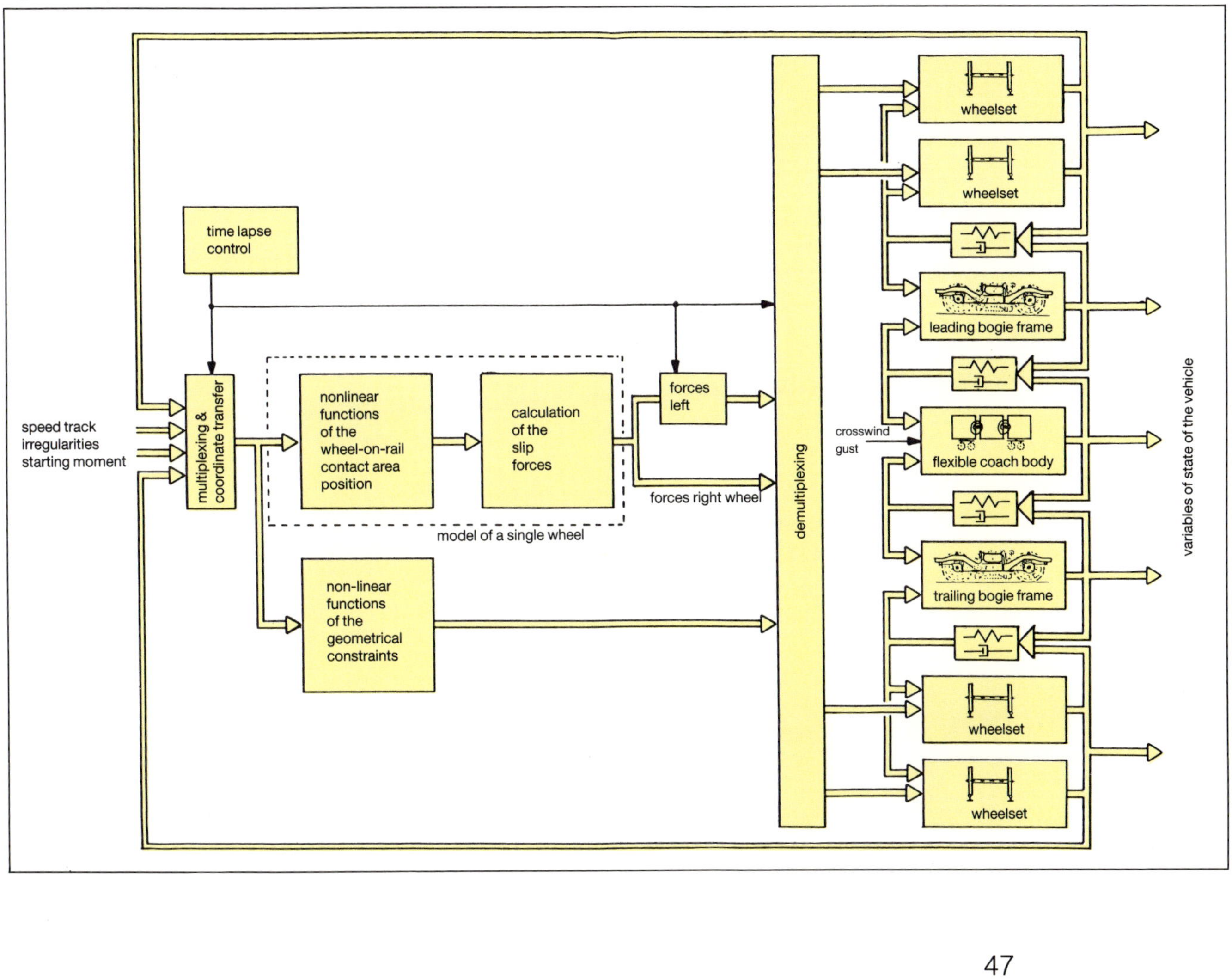

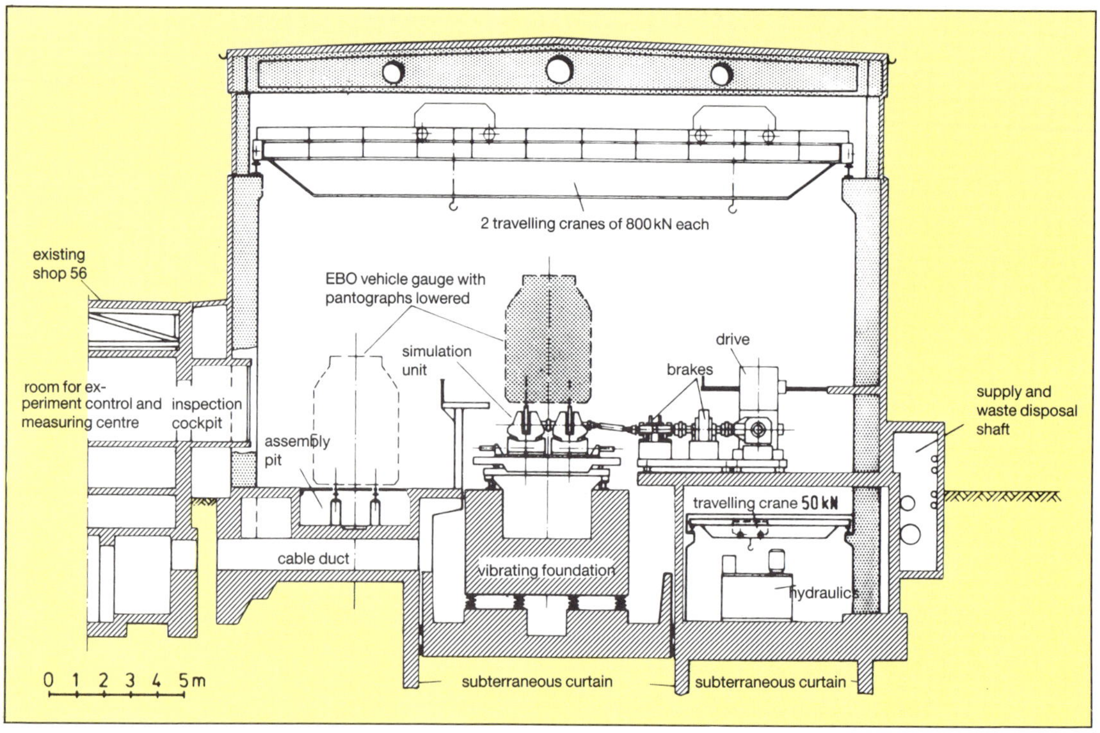

Fig. 5: Cross section through the test shop of the roller test rig

ing during running. Fig. 6 shows a two-axle unit of the roller test rig. The experiments are controlled and monitored through a central computer system, which also collects and evaluates the data. The control centre is shown in Fig. 7.

To-date mainly the following experimental setups were investigated on the roller test rig:

Experiments for the verification of the theory

The model sets used in the previously mentioned computer program were investigated with special experimental bogies, to check the agreement between theory and computation, to correct it if necessary, and to carry out a calibration of variable computational parameters.

Testing of a newly developed wheelset

In this experiment slip compensated wheelsets, in which the rigid axle was replaced by an adjustable moment transfer unit were investigated. Evidence of the fundamental function for the high-speed ranges was to be produced and the modulation concept optimized on the roller test rig. To do this two wheel-slip compensating wheelsets were installed in an experimental bogie of the roller test rig with adjustable primary and secondary suspension (Fig. 8). It was possible with these experiments to demonstrate this function up to a speed of 503 kmph.

Experiments with a newly-developed bogie component

Instead of the conventional hunting control an actively controlled anti-hunting device was installed in a bogie. The results of the theoretical calculation have already been presented in Fig. 2. The outcome of the investigation showed that with active hunting control the experimental bogie ran steadily throughout the entire speed range even when actuated by the rollers of the test rig, whereas without hunting control the critical wavy run occurred already at

Fig. 6: Basic cells of the roller test rig with drive and brake units

160 kmph. During subsequent trials in a vehicle speeds in excess of 500 kmph were attained on the test rig.

Checking the design data of a carrying bogie

During these tests the design data for bogies to be used in the ICE at up to 350 kmph were investigated to obtain information about the riding behaviour across faults in the track, the stability behaviour in the limiting region between the stable and the unstable state of riding and about the dynamic rigidities. Results were also produced for the structural design about materials stress, braking behaviour, and the effect of bearing clearances.

Investigations of the three-phase drive system of the electric locomotive BR 182 001

For testing purposes a three-phase drive system with intermediate circuit was installed in the locomotive 182 001 for the investigation of performance and function of this system on the roller test rig. This was the first power unit (Fig. 9) to be tested on this rig. The experiments served for the confirmation of the results produced through simulation during the design phase and through measurements during line trials. The main emphasis was on the investigation of the electrical drive system rather than on questions of riding quality. The test rig was used for driving and braking within the whole performance range. The measurements confirmed the results of the advance calculations.

Investigations of rubber-ring sprung wheels in the ET 473

The effect of these wheels, used in the Hamburg S-bahn coach ET 473, was investigated at various vibration excitations and compared with conventional wheelsets. The aim of this investigation was the reduction of vibration of buildings that can be caused by passing trains. It was possible to determine from the transfer function measured the frequency ranges within which the intermediate rubber layer becomes effective.

Trials of the DE 2500 "UmAn" experimental locomotive

A four-axle diesel-electric locomotive, DE 2500, was fitted with newly developed motor bogies and drive systems for the high-speed range of up to 350 kmph (Fig. 10). The special characteristic of this running gear development is the adjustable recoupling of the driving masses. In test rig experiments the effects of the various states of coupling on the specific behaviour of the experimental vehicle and the riding stability were investigated on various measuring runs. The motor bogies tested here are used in the ICE.

Basic investigations of a single wheel

Basic investigations were carried out for a Swedish research project on the roller test rig; it involved a vehicle with actively steerable individual wheels. The object of the experiment consisted of a cylindrical single wheel suspended in a frame so that it could be rotated round the vertical axis; the frame was mounted directly on a simulation unit which had de-

Fig. 7: Control centre of the roller test rig

Fig. 8: Experimental running gears with slip-compensated wheelsets

vices for changing the wheel load, for braking, and for the determination of the forces between wheel and frame. The play of forces between wheel and roller at various speeds and wheel contact area forces were investigated.

Checking the safety of loads during transport

The test rig permits optimum choice of load securing devices through the simulation of dynamic effects on the wagon loads.

The roller test rig represents the track and its inherent dynamics in a form diverging from reality. Corresponding experiments on the line and on the test rig were necessary for the conversion of the measurement results obtained on the rig into line conditions. Equally the theory verified for roller test rig conditions had also to be confirmed for the line. For this purpose an experimental vehicle was built, which incorporated in its centre an experimental bogy for the

Fig. 9: Locomotive 182 001 on the roller test rig

Fig. 10: Experimental locomotive DE 2500 on the roller test rig

necessary investigations; it is shown on the test rig in Fig. 11.

The roller test rig has proved to be a successful investment. Work to improve the control equipment is still in progress to make the behaviour of the rig resemble line conditions even more closely, so that the mathematical interpretation of the test results can be minimized.

For future use in the ICE, components and vehicle units are first tested on the rig and their behaviour is examined.

Vehicle research sector

In this research sector vehicle components are investigated for optimum design and layout. In this context not only the static and dynamic behaviour of the mechanical elements is considered, but also the potentialities of novel raw materials and use of alternative structural principles are investigated, for instance to achieve more favourable values of weight, which reduce power consumption.

The aerodynamic design of the individual vehicles and of the train set is partly responsible for the power consumption, noise generation, but also for the safety of the train at high speeds. Theoretical and experimental investigations were carried out into vehicle resistance, the design of the ends of the power units, and the forces of cross winds (Fig. 12). The effect of high-speed vehicles on passing trains, on buildings, and on people was also examined. Further research work was conducted into running in tunnel at high speeds, the design of tunnel entrances and

Fig. 11: Experimental vehicle 1 on the roller test rig

exits, and the problems arising from the generation of pressures. The results were incorporated in the layout and design of the ICE.

Another important subject on which work concentrated was the development of wheelsets for ultra-high speeds; in one case the design possibilities of the wheel disc as an elastic structure for the reduction of unsprung masses were examined. Fig. 13 shows this wheel in the centre; the other wheels demonstrate different structural principles for noise suppressors. The wheelset with elastic wheel disc is being tested on the line.

Running gears were developed with the aid of the already mentioned computing instruments, involving

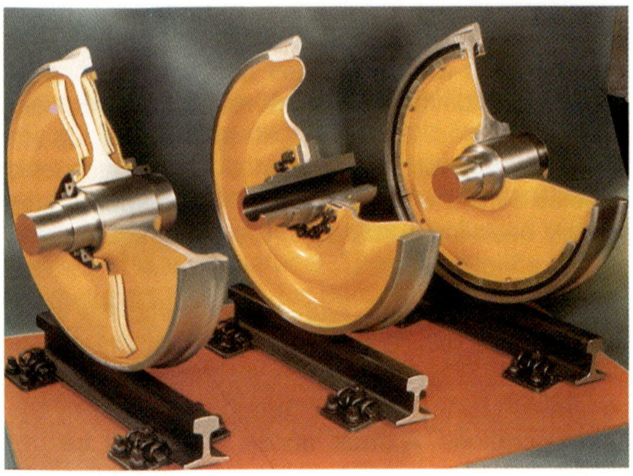

Fig. 13: Wheel disc designs

both motor and carrying bogies. The motor bogie designed for the ICE was tested both on the line and on the roller test rig (Fig. 10). An ICE carrying bogie with coupling frame was also investigated on the test rig. The construction principle is illustrated in Fig. 15.

The dynamic behaviour of the vehicle cells themselves was the subject of a number of theoretical and experimental investigations. Here, space- and

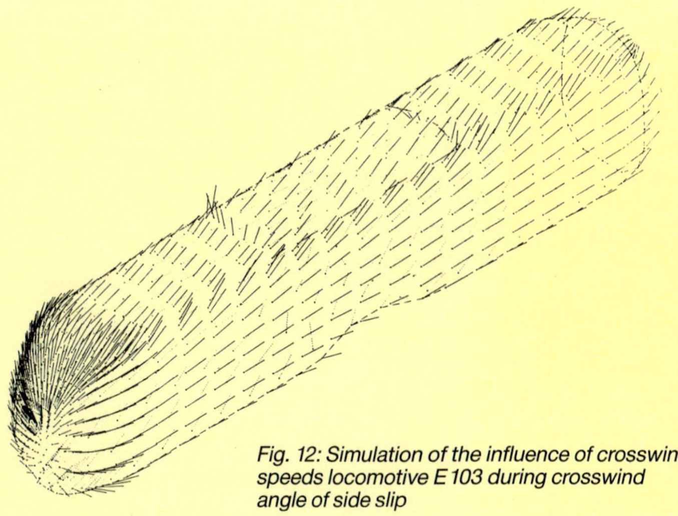

Fig. 12: Simulation of the influence of crosswinds speeds locomotive E 103 during crosswind angle of side slip

weight-saving structural solutions were the aim, which conformed to the demands of railway technology. Fig. 16 shows the result of a calculation of cell

Fig. 14: Coupling frame bogie

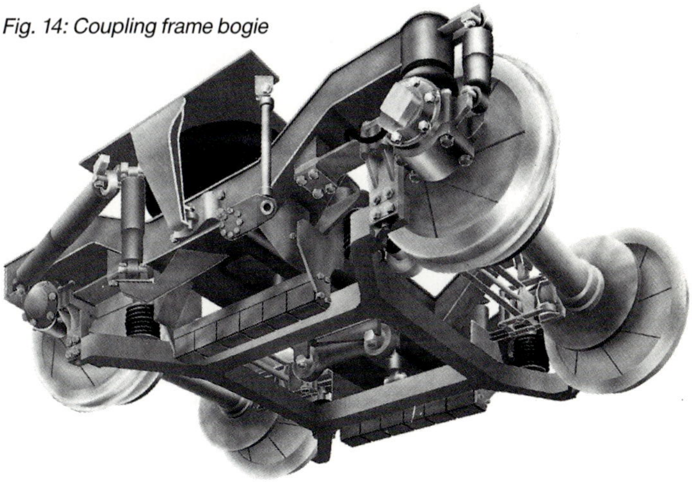

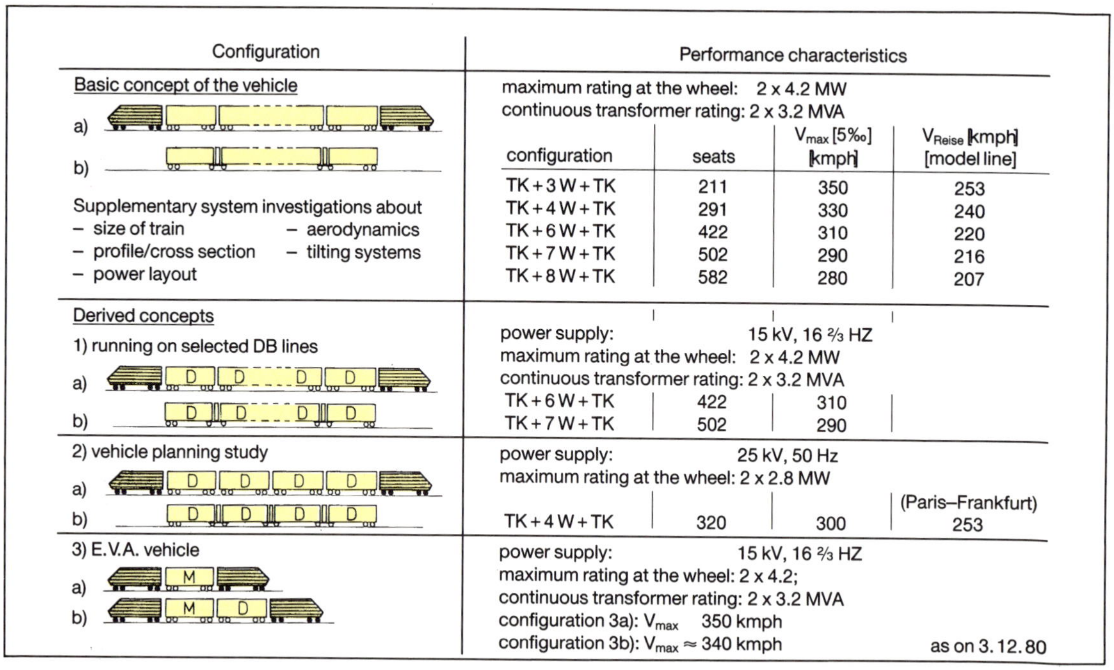

Fig. 4: Vehicle concept of the definition phase

tion available, and attainable speeds. Further possibilities, not described here, consist in the modification of the performance capacity of the power units or in the use of booster vehicles.

Power units of other current systems but using the same basic concept can also be derived. Thus, for the vehicle of the wheel-on-rail planning study – a joint Franco-German project in the Paris–Frankfurt/M planning region – a 25 kv/50 Hz power unit is envisaged which, with the exception of the modifications called for by the different current systems, will be identical with the ICE/V power unit.

For the concrete realization of the ICE/V, the variant shown in Fig. 4 has been chosen from the overall concept according to Fig. 5. The adoption of separate bogies is due to the conditions of the experimental line which, because of its limited length and in consideration of running speeds of up to 350 kmph, permits the use of only one, or at most, two centre trailers.

In March 1981, the result of the definition phase was presented as the first important milestone in the development of the ICE/V: its project description, also forming the basis for the tendering agreed by the BMFT, was announced in the Federal Gazette in February 1981.

At the request of the BMFT, DB carried out a detailed evaluation of the development proposals submitted by the industry.

After long-drawn-out funding negotiations and unequivocal comments by DB on the practical aspects of the ICE/V, the green light for the project was at long last given in September 1982. The German high-speed train was to be developed and built by the end of 1985 – the 150th anniversary of the German railways – within an agreed overall budget of DM 72 m.

Development and planning

The work on development and planning of the ICE/V began in 1982; it was even possible to base part of it on in-depth design calculations during the definition phase. The work concentrated on four main aspects:

▷ the mechanical part of the power units, which was the responsibility of a working party of Krupp (managing), Krauss Maffei, and Thyssen-Henschel;

▷ the mechanical part of the centre trailers including interior furnishings and air-conditioning plant, MBB managing;

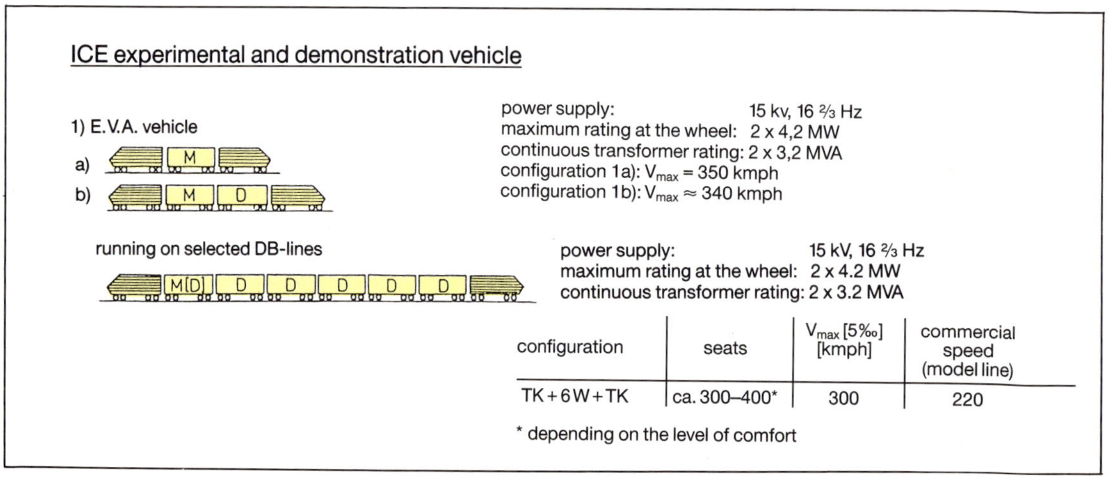

Fig. 5: Concept selected for RS-VD

▷ the electrical equipment for the power units and centre trailers in cooperation between Siemens and AEG, and

▷ the cross-section components of the train unit (braking equipment, diagnostics etc.) coordinated through the project management.

A special position was occupied by the measuring technology for the execution both of the comprehensive experimental programme and of the high-speed runs of newly developed vehicles, for which the monitoring of important safety-related parameters is indispensable.

Some aspects of the course of development are briefly discussed here.

Effects of the changed widths of the vehicles

The required unrestricted use of the ICE/V on the IC network of DB and the idea of a future transnational operation of high-speed multiple units, led to the adoption of the UIC vehicle gauge for the power units and centre trailers. The reduced width of the power units (3,070 instead of 3,200 mm) called for a rethinking of the ventilation concept and for a re-arrangement of the ventilation grids. The originally planned ventilation, through the engine compartment, was abandoned in favour of direct ventilation with lateral ventilation grids and air ducts to the air consumers, such as traction motor and oil radiator. In effect, this new concept appeared altogether superior to the original solution.

The KS 20 shape of the power units, too, had to be adjusted to the new width dimensions. To achieve a more pleasing appearance, minor alterations were introduced to the ends of the KS 20, within the scope of the aerodynamically permissible latitudes.

After thorough investigations it became clear that it was no longer possible to accommodate the originally planned bogie skirts on the centre trailers, now measured only 2,930 mm width, and with the given bogie design.

Development of the DSA 350 pantograph

The DSA 350 pantograph, newly developed by Dornier-System-GmbH (DS), Friedrichshafen, is an example of the development systematics applied to the ICE/V project.

Dynamic optimization

Optimization calculations based on earlier investigations were carried out for the interaction between the pantograph and the contact wire. For this purpose six spans of the Re 250 overhead contact wire system were taken into consideration and the behaviour of the system was observed during the running-in of the contact strip. After the dynamic behaviour of the overhead contact wire system had been determined, it was possible theoretically to describe, with the aid of the three-mass system, the pantograph and to couple it to an integrated system through the force of the contact strip against the overhead wire.

In addition to the masses and the springs damping for the individual systems was calculated and their effect on the overall behaviour was investigated. The finite-element method was used for the mathematical modelling. Accordingly the step-by-step integration of the movement equations was carried out for the part systems and an appropriate assumption iterated for the contact forces until the two systems were in equilibrium. It was necessary to investigate six spans of the overhead contact wire system and to calculate in conjunction with the 3-degrees-of-freedom model of the pantograph about 200 eigenmodes, within the frequency range between 0 and 22 Hz.

Qualitatively the result can be summarized as follows:

▷ mass of bow suspension as low as possible;

▷ bow suspension damping as little as possible;

▷ stiffness of bow suspension less than that of the frame.

The results of the calculation also show that with adherence to the specified limiting values, a maximum speed of 350 kmph could be obtained for the pantograph interacting with the Re 250 system.

Aerodynamic optimization

The demand for constant contact pressure of the pantograph against the overhead wire is expressed by the following criteria:

▷ low air resistance in the run-in state;

▷ low air resistance in operation;

▷ as identical as possible aerodynamic behaviour with different incident blow (knuckle leading and knuckle trailing direction);

Fig. 6: DSA 350 pantograph on E 103 locomotive

▷ low turbulence demand;

▷ contact pressure against the overhead wire less than 120 Newton, if possible;

▷ eigenfrequencies of the pantograph and of the contact wire must not be affected by generated vortex frequencies.

Because the geometry of the contact strip greatly influences the aerodynamic behaviour, model investigations (2.5:1 scale) were carried out at the Berlin Institute of Turbulence Research of the German Research and Experimental Institution for Aeronautics and Space Travel into the effect of the shape of the contact strip. It was possible to demonstrate that the rectangular shape met the requirements most effectively.

The results of the theoretical simulations and experimental investigations were applied to the layout and construction of the pantograph. After completion of the prototypes, trial runs at up to 280 kmph were carried out (Fig. 6).

Passenger information system

Beside the on-board diagnostic system and the measuring plant for monitoring and experimental purposes, the passenger information system in the ICE/V is the third component for which a detailed definition was required during the planning period, because at the beginning of the work it had not yet been possible to describe the requirements with sufficient accuracy.

Fig. 7: Standard seat module for 1st Class seat

The ICE/V multiple unit offered the possibility for providing the passenger with information systems expanded beyond the present scope, for testing passenger reactions during demonstrations, and for obtaining information on acceptance or rejection. Knowledge about acceptance of wide-ranging possibilities of information thus acquired will enable the operator to decide on the commercial application of the systems of passenger information offered on the ICE/V.

The nucleus of the installation is the computer in the train centre, which in the ICE/V is housed in the dynamometer car. The central computer is connected with the individual information facilities of the train through a wide-band coaxial train cable, which also passes through the automatic Scharfenberg coupling.

The production train, too, will be equipped with a comprehensive passenger information system, offering communication and entertainment to be described later. Figs. 7 and 8 show examples of two solutions for the ICE/V armrest module, with the controls for radio and reading light and the at-seat communication centre with paycard telephone and monitor screen.

Second demonstration coach

In March 1983, the Board of Management of DB decided to have a third centre trailer for the ICE/V developed and built, on their own account, to extend the possibilities of choice for future production runs. At the same time, the now five-car ICE/V would approach the character of a full-length train more closely and thus enhance the demonstration effect.

The additional third coach follows the design principle of the first two coaches, but in execution differs in the following features (Fig. 9):

Fig. 8: Communication centre with video screen and on-board telephone for 1st Class seat

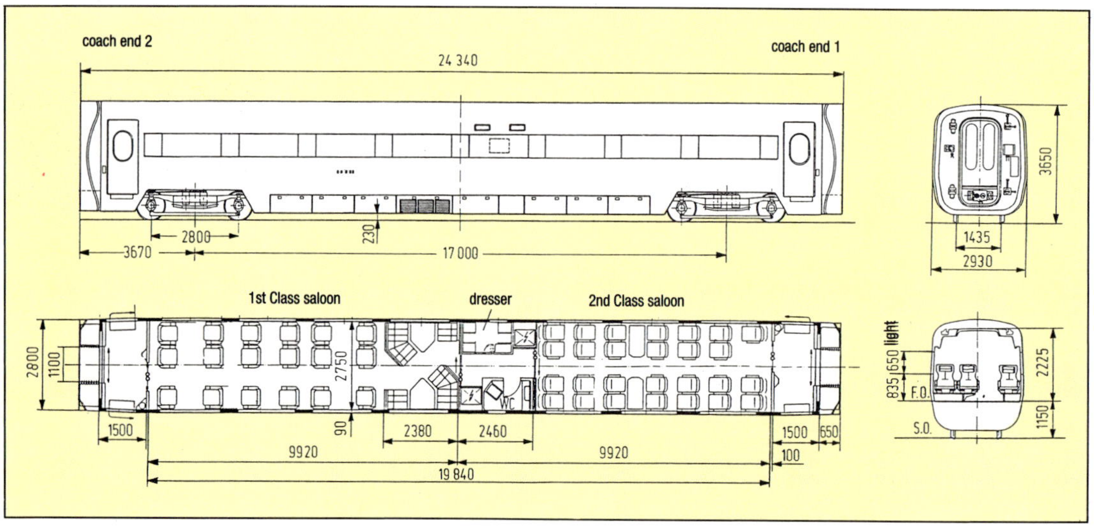

Fig. 9: Ground plan of the 2nd demonstration coach

Fig. 10: Lowering the power unit onto the bogies

▷ centrally located technical, W. C. and washroom facilities (central block section);

▷ two passenger compartments separated by the central block section; "see-through" vestibule design with glass elements;

▷ layout as mixed 1st/2nd Class variant with conference compartment;

▷ snack and bar service for demonstration runs.

Although ordered much later, the third coach was delivered at almost the same time as the first two coaches.

Progress in construction

At the beginning of 1984, the first cut-to-size sheets for the construction of the two power unit body shells were produced by Krupp, Essen.

The first completed car body was subjected to the regulation pressure test by the experimental establishment of the Federal Railway at Minden/Westphalia.

During the summer of 1984, the colour scheme was passed by the management of DB in time for the painting work to begin.

The outward appearance of the ICE/V is influenced largely by the aerodynamic design of the power units and by the continuous window strip of the centre trailers. The livery has been designed subtly to enhance these dominant elements. For the ICE/V, a top coat in grey-white with quartz-grey skirting was chosen. A continuous red-violet longitudinal band emphasizes the unit character and the "elongated" appearance of the multiple unit.

From October 1984 onwards, the power units were ready for the installation of the electrical equipment. In January 1985, they were lowered onto the motor bogies for the first time (Fig. 10).

At the end of 1984, the installation of the equipment also began in the centre trailers.

The new travelling and resting seats for both demonstration coaches were developed by Grammer Sitzsysteme, Amberg. The first batch of 1st Class seats is finished in the modern foam-back technique with adjustable backs and two armrests independent of the neighbouring seat. Every seat has a hinged table and is prepared for the installation of the armrest module for the passenger information system.

The 2nd Class seats developed for the second demonstration coach, from July 1984 onwards, have also been produced according to the foam back principle. Seat kinetics and ergonometric shaping of the upholstery are the same as those of the 1st Class seat, except that the 2nd Class standard had to be conformed to, ie narrower width of seats (four instead of three seats in a row), restricted angle of tilt of the back because of the narrower seat divider.

Handing over to DB

On 19th March 1985, the first power unit, Serial No 410 001, was accepted by DB on the occasion of a joint inauguration ceremony, sponsored by Krupp and DB, in the presence of leading representatives of the Federal Ministries of Research and Technology and of Transport.

On 31st July 1985, the time had arrived for the centre trailers at Donauwoerth to be ceremoniously handed over to DB; on this occasion the power unit 410 001, too, was transferred from Freimann to Donauwoerth (travelling, by the way, under its own power), so that for the first time the impression of the "integrated train" could be conveyed to a large number of invited guests (Fig. 11).

Fig. 11: Handing over of the ICE/V at Donauwoerth

By the beginning of September 1985, all the centre trailers had reached Freimann to be marshalled into the train with the power units.

After the completion of the initial operation and acceptance runs, which took place in the Munich region until the end of September, and subsequently on the Brackwede-Hamm high-speed line, the ICE/V was introduced to the public by the management of DB, on 26th November 1985, on the occasion of a special trip from Bielefeld to Essen, when at 317 kmph the 300 kmph "barrier" was clearly breached for the first time in Germany.

This concluded the first phase of the ICE/V project. The second phase followed from January 1986: detailed trials of the ICE/V for the assurance of the production development. The results of these tests will be reported in detail elsewhere.

From the IC-Experimental to the InterCityExpress (ICE)

As far back as 1984, preparatory work had begun on the InterCityExpress (ICE) high-speed production train, which from May 1991 onwards is to run at speeds of up to 250 kmph on those new and upgraded lines which have been completed by then.

After agreement on the specifications of the production train had been reached, a preliminary specification with an outline production schedule was submitted at the beginning of 1985.

After a thorough investigation of numerous details and of individual components, the specification for the ICE was submitted on 2nd January 1986; it formed the basis for the tender and for the award of the orders for the production trains.

Overall concept of the ICE

The ICE, which will be given the Class designation 401/801 by DB, adopts its basic concept of the multiple unit from its forerunner, the IC Experimental, designated 410/810 by DB. The basis of development was abandoned only where commercial requirements, technological aspects (also derived from past trials), and economic considerations called for modifications.

DB's ICE production train is therefore characterized by the following concept features:

▷ multiple unit, consisting of two power units with three-phase power transmission (intermediate voltage circuit and asynchronous traction motors) and centre trailers;

▷ performance layout for 14 centre trailers and 250 kmph service speed;

▷ traction and braking equipment suitable for 280 kmph; development potential installed for up to 300 kmph;

▷ splitting of the multiple units on repair tracks only; no quick uncoupling of the individual vehicles;

▷ aluminium centre trailers in large extruded profile construction, with continuous window strip;

▷ maximum width of the centre trailers specified at 3,020 mm at a length of 26.4 m for improved seating comfort, especially in 2nd Class;

▷ uniform roof height of 3,840 mm above top of rail for power units and passenger coaches;

▷ service coaches with vaulted roof and additional fanlight windows;

▷ aerodynamically optimized design as with the IC Experimental.

The operationally integral multiple unit will have all the decisive components laid out for a speed potential of up to 300 kmph, ie running gear, pantographs, and braking concepts conform to the outline specifications of a European 300 kmph multiple unit without the need for new development.

The well-tried three-phase drive technology has made decisive progress even during the development period of the ICE in digital control technique, in the power converters with GTO elements, and in converter cooling. Some of the ICE trains will already benefit from this advanced technology in the first power unit.

Fig. 12: Intercommunication between coaches with double rubber joints

The construction of the production centre trailers also using the aluminium extruded profile method has led to the adoption of a closely integrated project procedure by the semi-finished products manufacturers, wagon builders, and paint manufacturers to ensure the success of this construction method. The improvements already introduced in the ICE/V centre trailers from known developments are fully accepted here. Attention is paid to aluminium-friendly construction, to anticorrosion protection, and to paint adhesion. Expected advantages of the use of relatively thick-walled extruded aluminium profiles, which can be produced at up to 800 mm circumference diameter are:

▷ low production costs;

▷ a reduction of dynamically sensitive welding seam lengths;

▷ increased resistance to the loading condition of "stress caused by pressure change during the meeting of passing trains in tunnel".

The experimentally proven good aerodynamic and aeroacoustic properties of the ICE/V are to be maintained in the ICE. A careful detail optimization in the course of development of the ICE is therefore necessary to meet these requirements, because some changes from the forerunner vehicles have become necessary:

Fig. 13: Shape of the ICE-power unit

▷ mounting of a second pantograph for running in Switzerland (can be partially compensated through moving the master switch under the roof into the engine compartment);

▷ replacement of the flush outer shroud of the intercommunicating gangway of the ICE/V with the least costly solution, consisting of a disconnectable double rubber joint and external wind-deflecting profiles (Fig. 12).

Design of the power units

The power units of the ICE are described in detail elsewhere in this book; here only a few points are mentioned for the better understanding of the overall concept of the ICE.

The concept of the power units largely follows that of the forerunner trains. The slightly altered cross section of the centre trailers, with only one side wall break at 1,800 mm above top of rail and the bottom side wall slope reduced from 10° to 5.2° to achieve improved boarding conditions for the centre trailers, calls for a minor modification of the shape of the power units to maintain the vehicle profile; thereby it appears to be slightly slimmer (Fig. 13), although it continues to display the basis of the development.

The major changes from the ET 410 are as follows:

▷ continuous rating increased to 2 × 4,800 kW to meet the train haulage programme;

▷ enlargement of the driver's compartment and further development of the driver's console with the inclusion of modern display technologies (Fig. 14);

Fig. 14: ICE-drivers desk with modern indicating instruments

▷ omission of the continuous 15 kv cable between the two power units, because with the projected running speed and length of train the pantographs of both power units can make contact with the overhead wire and obtain thus power;

▷ two cast steel brake discs instead of three for each axle in the otherwise unchanged motor bogie; omission of the eddy current brake from the motor bogie;

▷ improved pressure sealing of the driver's compartments.

Control technology and diagnostics

The principle of the control technology of the ICE (Fig. 15) is based on the establishment of a redundant system of reasonable scope, which, for instance, permits the stopping of the vehicle when an isolated breakdown occurs, or, in the event of a complete breakdown of the traction part of the leading power unit, assures its function as the control unit for the still functioning trailing power unit. The fibre optic technology of the ICE/V for the in-train data bus is retained.

The diagnostic system provided for the ICE is hierarchically structured. It operates independently of the control and regulating functions; interference with the vehicle functions is not envisaged. The diagnostic system will work:

▷ before the start of a run as stationary test or as an automatic preparation routine;

▷ as function diagnosis during running;

▷ as maintenance diagnosis for the repair shop.

The diagnostic structure is characterized by the following features:

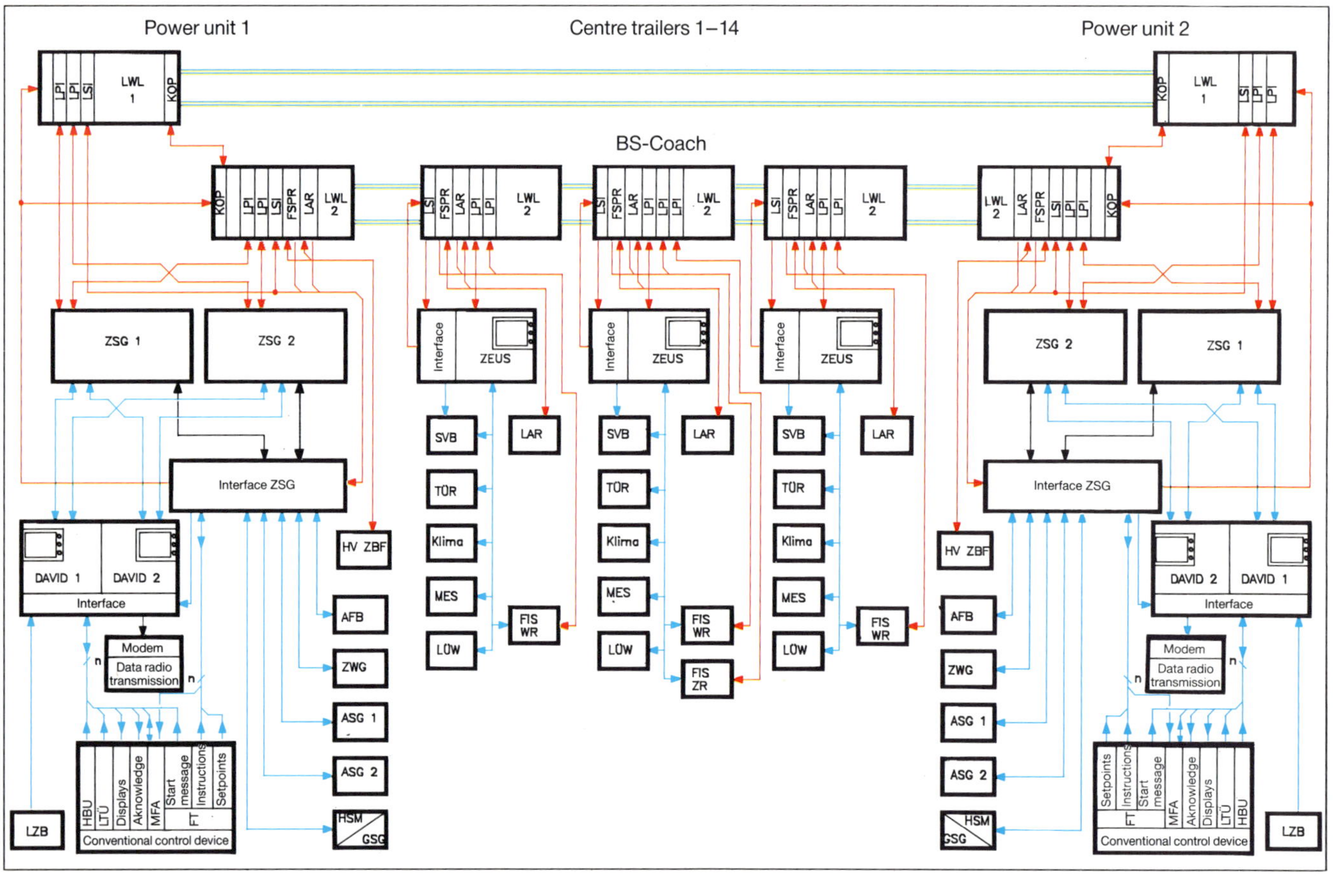

Fig. 15: Control technology of the ICE, demonstrated with the example of the power unit, in combination with the diagnostic computer

Legend:

ASG	Drive control
ZSG	Train control
LZB	Continuous automatic train control
HVZBF	Main distributor ground train radio
HSM	Type designation for driver's brake valve appliance
GSG	Wheel slide protection
MFA	Modular driver's cab display system
HBU	Auxiliaries converter
LTÜ	Bearing temperature monitor
ZWG	Central distance and speed indicator
FT	Driver's console
AFB	Automatic traction- and brake control
LWL	Fibre optics station
LPI	Fibre optics processor interface
LSI	Fibre optics signal interface
FSPR	Interface telephone
KOP	LWL coupler
LAR	Interface loudspeaker installations in the passenger coach
DAVID	Diagnostic and preparation service with integrated display control
ZEUS	Central installation for monitoring and control
SVB	Signal processing brake
FIS	Passenger information system (audio/video system, telephone with C-network)
MES	Measured-value indication system
LÜW	Running state monitor
KLIMA	air-conditioning plant
TÜR	door control unit

▷ subsystem-diagnosis for the important components of power units and centre trailers as far as these components are of suitable design and can be fitted with suitable interfaces;

▷ on-board data concentration in the vehicle computer (diagnostic and preparation service with integrated display control (DAVID) in the power unit, central installation for monitoring and control (ZEUS) in the centre trailer), with man/machine interface (display and keyboard). The vehicle data displayed are on "operational data" priority in the power units, and on "train crew and inspector service" priority in the centre trailers;

▷ diagnostic up to the plug-in card as far as possible and technically practicable. Where this can only be achieved with unacceptable outlay, at least go/no go information on the components or subcomponents is aimed at;

▷ data links between subsystem and vehicle diagnostic instruments;

▷ in the power unit as serial link with 20 mA interface (as in the IC Experimental, experience of which is being utilized here);

▷ in the centre trailer as RS 485- serial vehicle bus;

▷ train diagnosis with data link via the fibre optic cable of the train control. Agreed-upon diagnostic signals of the train set (for instance for brake control, door control etc.) are filed in the store of the DAVID power unit appliance and – as far as required for operational control – displayed on the driver's compartment display or, if required only, for maintenance collected for future data output. The DAVID instrument is therefore both power unit computer and central train computer;

▷ remote data transfer to a centre in order to keep as short as possible "dead" times for maintenance.

Layout of the centre trailers

The following conceptual aspects apply to the ICE centre trailers:

the 1st and 2nd Class coaches constructed are with different window dividers so that the desired levels of comfort can be achieved. The interior layout of both classes of coach will be of modular construction and in one coach offer three different facilities to the passenger:

▷ compartments of classic concept with a 775 mm wide side corridor and compartment doors;

▷ open saloon with rows or groups of seats or individual seats, separated by wardrobes; rest seats developed for the ICE with 40° reclining backrests;

▷ open saloon with facing seats and tables, separated by wardrobes.

The seat rails will be continuous, even in the compartment areas, so that, at least in the open saloon areas, a change of the seat divider will be possible to accommodate changing market demands. For reasons of cost, more far-reaching preparations for the conversion of coaches from the compartment to the open saloon version have not been implemented.

In the interest of the passengers, there will be a strict division between smoking and non-smoking coaches in the ICE. Standard comfort includes draught-free air conditioning with ceiling ventilation and underfloor heating, retention toilet systems operating on the vacuum principle, and two paycard telephones, to be housed in a booth each in the 1st Class and 2nd Class section of the train. In the conference compartment a third telephone with call charge printer and mobile handset will be available as an additional service facility; it also permits faxing.

In view of the large number of tunnels on the new lines

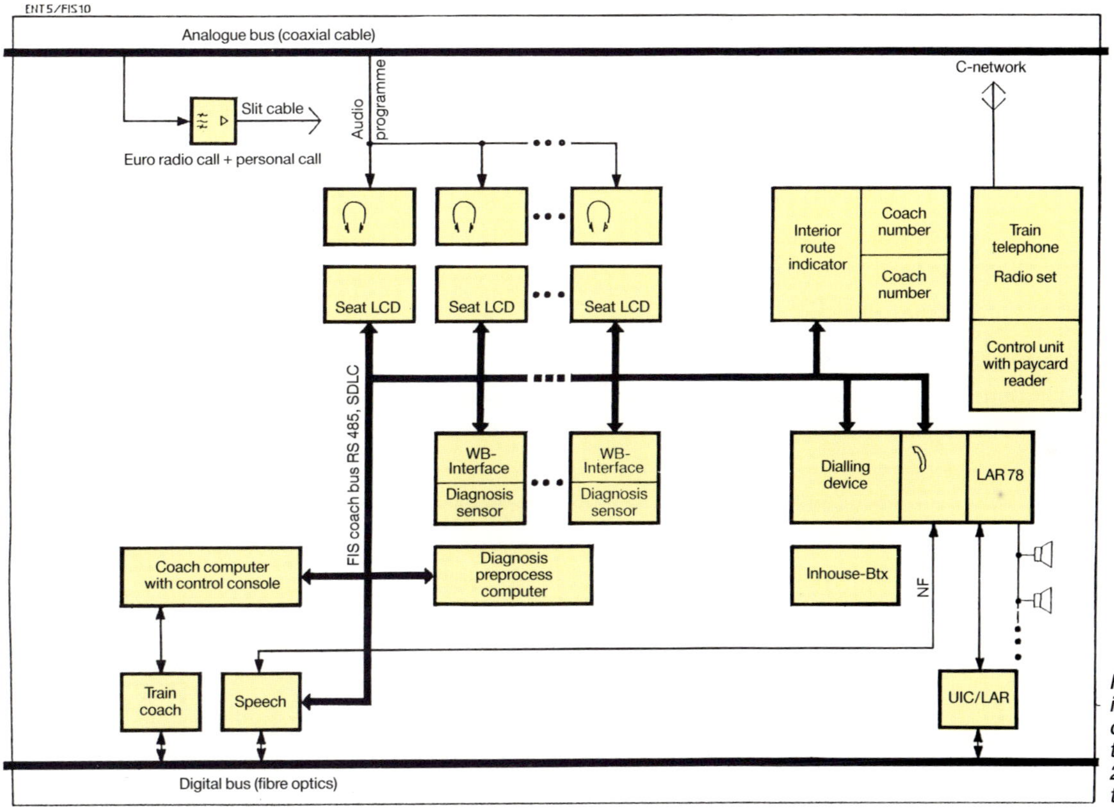

Fig. 16: Passenger information system, demonstrated with the example of the 2nd Class coach with the control room

of DB, the air conditioning plant will be pressure protected to attenuate the pressure shock waves generated inside the coaches by passing trains, especially in tunnel, so that passenger comfort is not affected.

The first batch of centre trailers currently being built will have steel-sprung bogies of the MD 530 version, with double mechanical hunting control and four brake discs. For subsequent batches the air-sprung bogie as fitted to, for instance, the forerunner train of the ICE, will be developed to operational maturity.

As in the forerunner, ET 410/810, the centre trailers are supplied with power by two 1000 V train bus bars (each supplied by one power unit) and a 110 V battery bus bar.

Service coaches

The service coach is, as usual, positioned between the 1st Class and the 2nd Class rake. The kitchen block is centrally situated and segregates the restaurant of superior comfort level from the self-service snack area.

During heavy demand for restaurant service, it is intended to serve meals in the adjacent 1st Class coach, in the area with facing seats and tables.

To make the service coaches conspicuous in the train set they have vaulted roofs, raised to 4,295 mm above top of rail and sloping at both ends.

Passenger information system

Like the ICE Experimental, the ICE will be equipped with a passenger information system to include initially the following services (Fig. 16):

▷ interior coach display of seat reservations and information;

▷ public address system;

▷ conductor call system (service call);

▷ European radio call;

▷ at-seat audio programme;

▷ video facility in the open saloon areas of the end coaches (1st and 2nd Class);

▷ on-train cable text system (in-house Btx) with mail box function.

The passenger information system will also use the train data bus; for the wide-band facilities (audio/video) a coaxial cable will be installed. For the operation of the playback and control units, an operating centre will be installed in the attendant's compartment. This centre (Fig. 17) contains the central passenger information system computer with operating panel and printer as well as the playback and reception units for audio and video programmes and UHF radio transmission. For regularly recurring announcements (standard announcements), a playout message unit is available, controlled, depending on its location, by the central passenger information system computer.

Certain details of the audio/video facilities are worth a special mention.

All seats have an audio module in the left-hand armrest with channel selector, volume control and headphone socket. Three radio-, three on-board, and two video programmes (audio) of stereo quality are available at the seats, received through electrodynamic headphones with conventional plug connections; reception in 1st and 2nd Class coaches is provided.

The backrests of the swivelling open-saloon seats contain a video module in LCD technology. Provision has been made for the installation of larger screens, to be available shortly. Because the seats swivel,

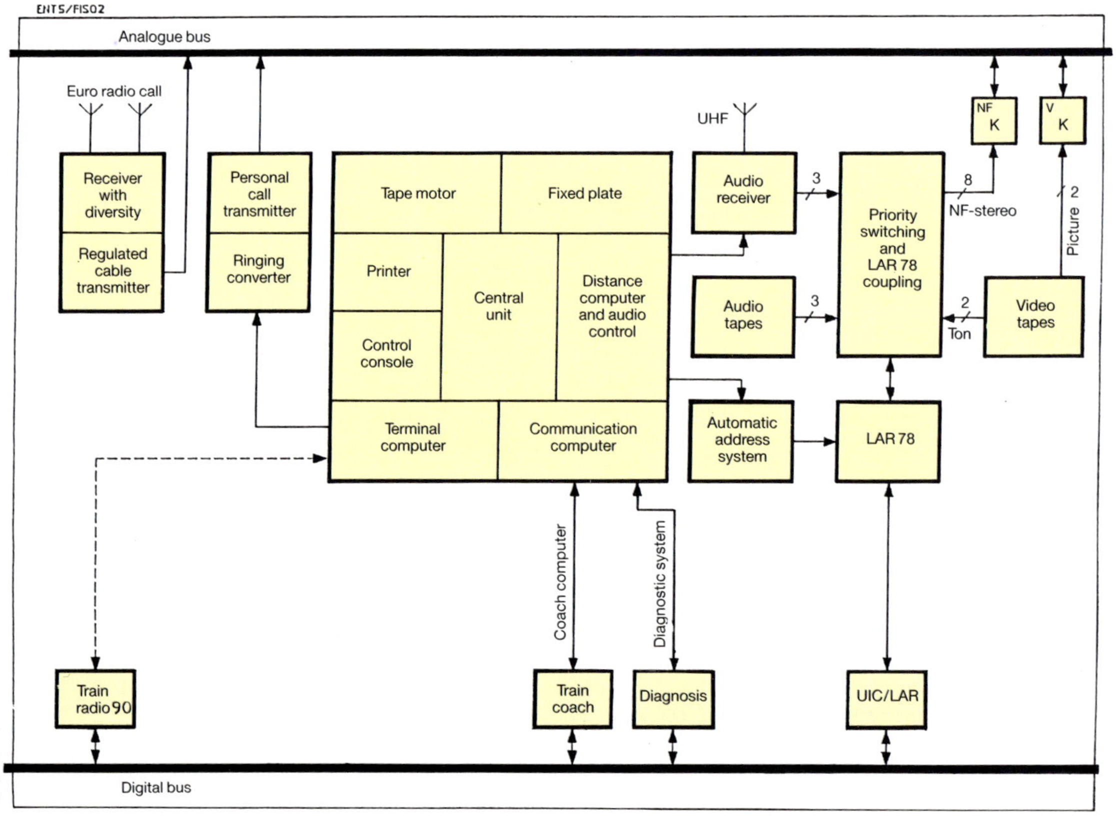

Fig. 17: Passenger information system components in the control room

there is no fixed relation between video screen and seat; video sound is therefore transmitted via the seat-related audio module. The LCD flat screen is mounted in a swivelling frame so that the changing inclination of the backrest can be compensated for. Two video programmes are provided, through which, for instance, service information can be fed in. Initially it is intended to equip one 1st Class and one 2nd Class coach each with the full passenger information system; the rest of the coaches will have basic equipment installed for retrofitting at a later stage. Each mini video screen has two controls for programme and brightness. The volume control is integrated in the armrest module.

For the further improvement of customer service, the passenger will soon have the facility of the service call.

A call button for the passenger (no two-way speech facility) is provided for each group of seats and for each compartment in the 1st Class coach. For the 2nd Class coaches the service call facility has been prepared, but for the time being it will not be operational.

The call buttons have an optical answering device through a two-wire bus. Communication between the train crew and the service coach is possible through the call station in each coach.

Despite all the doubts with which the equipment of modern trains with improved communication systems can be viewed, it must be borne in mind that these technologies are invading an ever expanding area of everyday life and are increasingly taken for granted as part of personal comfort.

The railway would without doubt suffer if in the year 2000 its most modern trains presented a glaring gap in communication facilities.

Delivery schedule

A first ICE line is planned, covering both the DB's new lines in June 1991. This called for the delivery in autumn 1989 – as scheduled – of the first power units (Fig. 18) to DB by the manufacturers. After the necessary initial operation procedures the first power units can commence trial operations from autumn 1990 onwards. The period of about 6 months left until May 1991 must be used intensely to iron out any deficiencies and weak points before revenue-earning service can begin.

Fig. 18: ICExpress power units being assembled

speeds of up to 400 kmph (Fig. 13), these loads are only around 1/3 of the permissible Prud'homme limit value.

In curves on old tracks (Fig. 14), at most only around 80% of the limit values was reached with a free transverse acceleration of 0.85 m/s². On new lines, the ICE/V travelled at 400 kmph with an unbalanced transverse acceleration of 1.4 m/s² (Fig. 15).

The dynamic vertical forces are also very low. On new lines at 400 kmph, max. wheel forces of only 112 kN were reached. In curves with 1.4 m/s² unbalanced transverse acceleration, values of max. 140 kN were recorded at speeds of 400 kmph. Measurements thus fell far short of the internationally applicable limit of max. 170 kN for wheel force even at high speeds and despite an axle load of almost 20 t.

The above test results confirmed that there was no need to modify the design for running properties used on the ICE/V when constructing the ICE. However, it was recognised that at very high speeds, specially-designed jaw dampers are required to damp small, low-frequency motion.

The running properties of the ICE bogies have been designed for high running stability. Safe running is ensured even when the limit values for wear to the wheel profile and rail have been reached or when individual dampers fail. Even with the unbalanced transverse acceleration occurring in present-day operations on the old lines still in use, the forces involved are well below the permissible limits. For further applications, e.g. travel with higher transverse acceleration values, simple variations to this bogie design can be made and have already been successfully tested on the ICE/V.

Structural design of the motor bogies

The bogie frame is designed as an O-shaped component group with no central cross member. This simple design has only 4 measuring points on the links of the axle guides; the relative distances between these measuring points can be checked with a high degree of accuracy.

Suspension – both primary and secondary – is provided in the form of coil springs. The secondary springs are arranged in the middle of the longitudinal bogie members to avoid torsional stress in these members.

Having axle guidance via one-sided rods with spherical rubber bearings and the bogie-to-body connection via a low-lying pull-push rod with large-volume elastomer joints, minimises the number of joints required for bogie guiding to 10 per bogie. All joints are wear and maintenance-free, since their motion is provided exclusively by the elasticity of the elastomer layers.

ICE braking system

In addition to the powerful regenerative brake, the ICE power unit is fitted with an electro-pneumatically operated compressed-air brake, in accordance with UIC standards. This brake acts on the mechanical brake in the power units and controls the brake system for the entire train – including the regenerative brake – via a computer-assisted brake control unit. On the basis of the deceleration selected by the power unit driver, the brake control unit activates the various brakes in the power unit and cars in the preselected sequence. Priority is always given to the wear-free regenerative brake in the power units.

Brake commands are transmitted electrically from the driver's brake valve to the brake control system. An emergency brake valve permits direct action on the train's compressed-air brake. Emergency stop commands emanating from the train can be bridged by the driver to allow a more suitable spot for emergency braking to be reached.

The brake can also be operated by an automatic train stop system (Indusi), a track-to-train communication and control system (LZB) and by using SBB's Integra System.

Fig. 16: Air frame with screw compressor and air drier

The compressed air required for the brakes and ancillary consumers is produced by a screw compressor with series-connected air-drying unit in each power unit (Fig. 16). The brake control units and virtually all other compressed-air appliances, such as stop cocks, solenoid valves, pressure monitors etc. are all included on a clearly laid out and easily removable braking instrument panel. The longitudinal compressed air pipes run along the central gangway beneath the running board, which means that the underside of the locomotive body is to a large extent free. The main air pipe and main air tank line are automatically connected to the centre trailers via the couplings.

Each axle has 2 cast-steel brake discs integrated in the drive. The solid discs have no internal ventilation and are fixed to the hollow axle using a pressure

Fig. 17: Mechanical brake in motor bogie

bond. The brake shoes with their sintered-metal linings are also fixed to the drive block and are operated by a brake cylinder (Fig. 17) on each axle. All brake cylinders are spring-loaded, allowing the entire train to be held safely at a standstill on gradients of up to 12.5% without applying the handbrakes on the centre trailers.

The ICE – the train for Europe

From the beginning of October 1989 onwards, the series-produced ICE power units are being delivered to the German Federal Railway at a rate of around one per week (Fig. 18). After each power unit has been put through a commissioning phase lasting several weeks, a sufficient number will be available to provide high-speed transport (250 kmph) on the new lines, which will be completed in time for the 1991 summer timetable. Thanks to the futuristic design of these vehicles, the next step of raising speeds to 280 kmph will present no technical difficulties whatsoever. Furthermore, there is the potential for development towards even greater speeds.

The first of these developments will be the ICE-M, a multisystem train capable of running on all standard-gauge systems currently used in Europe. This train will be designed for service speeds of 300 kmph on AC networks and 220 kmph on DC networks. On the basis of international agreements, the static axle load will be limited to 17 t. Although these demands are greater than those placed on the series-produced ICE, they can be achieved without any major modification to the basic ICE design.

ICE engineering thus offers a train suitable for use internationally, which will become increasingly necessary as European high-speed lines merge to form an extensive high-speed rail network; it seems unlikely that demands made on vehicles using this network will be fully homogenised in the foreseeable future. ICE high-speed trains will not only offer new standards of public long-distance passenger transport in Germany, but will also represent a high-tech solution to the most varied of requirements on an international level.

Fig. 18: ICE power unit, class 401

Klaus Becker-Lindhorst

The ICE brake equipment

Introduction

The brake on the ICExpress is based on that of the IC-Experimental and demonstration train, whose brake system was developed and produced by Knorr Bremse AG, within the framework of the wheel-on-rail research programme promoted by the Federal Ministry of Research and Technology.

Experience and knowledge gained from the experimental train led to further developments and improvements. Modifications were the result of some new specifications.

To achieve adequate brake performance in high-speed trains several mutually optimized brake systems must be used.

The brake equipment of the ICExpress (ICE) comprises:

▷ the electrodynamic brake of the asynchronous three-phase drive with grid feedback;

▷ two non-ventilated axle-mounted brake discs per carrying axle, four ventilated axle-mounted brake discs per trailer axle;

▷ electro-magnetic rail brakes in the carrying bogies with the option of eddy-current brakes.

Coach construction and brake control are designed so that instead of the electro-magnetic rail brake the eddy-current brake with controlled current supply can be installed. Whether or not the ICE production trains will be fitted with eddy-current brakes, will be decided after further investigations.

The objective and the concept of the layout of the brake is an operational braking action which is as wearfree as possible. The regenerative brake of the drive therefore has priority; if necessary it is assisted by the disc brake.

The wearfree and easily controlled eddy-current brake offers great advantages. For the ICE version with this type of brake, the overall brake system has been designed so that at decelerations of up to 0.5 m/sec^2 (maximum service braking with automatic train control) the train is braked solely by completely wearfree electric brakes (Fig. 1). Only in the lowest speed range, when because of the downward slope of its characteristic curve the eddy-current brake is switched off, must the disc brake be used. The resultant wear on the discs and on the brake linings is, however, insignificant in this speed range with low kinetic energies.

Disc brake

The disc brake takes over, on the one hand, the function of the service brake, and on the other, the task of a safety system upon emergency brake application, when the minimum braking distance must be safely maintained even during partial failure of the braking equipment.

Disc brake in the power unit

Although the power units are fitted with extremely powerful regenerative brakes (4 MW brake power per power unit) with grid feedback, these brakes are occurs out of action when failure occurs through failure of the contact wire. The dimensions of the disc brake were therefore based on the "worst case", ie braking to standstill with total failure of the regenerative brake.

In comparison to the IC-Experimental (ICE/V), each driving axle has only two instead of three brake

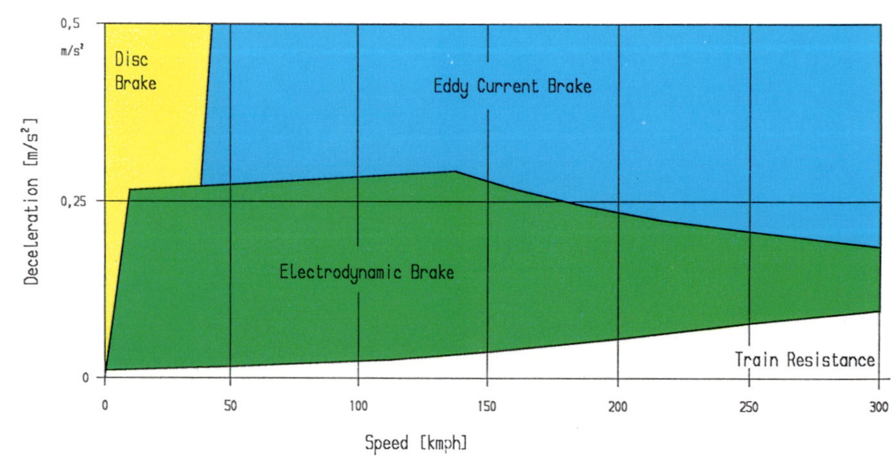

Fig. 1: ICExpress, version with eddy-current brake Interaction of the brakes

discs mounted on it for reasons of space, cost, and the increased number of centre trailers in the train set, which can assist in the braking of the power units.

Because the discs are rarely operated, only during failure of the regenerative brake and with braking above 160 kmph, it was possible to choose non-ventilated discs offering the advantage of minimum ventilation dissipation. Owing to the high thermal stress on the discs, a decision was made in favour of cast alloy steel brake discs with sinter metal linings. Such brake discs – a new development by Knorr Bremse AG – have been found reliable both in the ICE/V and in the TGV.

Fig. 2: Motor bogie with disc brake

Fig. 2 illustrates the motor bogie with the two solid discs (thickness 40 mm) mounted on the hollow axle. The brake force is generated via a double jaw by a 12″ brake cylinder with combined compressed-air and spring partion. The latter operates as a parking brake. With the spring brakes of the power units, a train with 14 centre trailers can be held safely at a maximum down gradient of 12.5‰ (1:80).

To increase the wheel/rail adhesion coefficient on slippery rails, the power units are fitted with electrically controlled sand metering units, which are thermostatically heated to keep the sand dry. The quantity of sand dispensed by these Knorr units is controlled in two stages, depending on the speed of the train, so that approximately the same amount of sand is applied per unit distance covered at high as well as at low speeds. The leading bogie wheelsets of the power units are always sanded.

Disc brake in the centre trailer

The overall concept of the brake presented a problem: how much consideration to be given to different requirements between the first ICE production batch entering service in 1991 and a future second batch [1].

ICE, first batch:

▷ maximum service speed 250 kmph;

▷ traction and braking designed for 280 kmph maximum speed;

▷ new Hanover–Wuerzburg line with 12.5‰ maximum down gradient.

ICE, second batch:

▷ maximum service and design speed 300 kmph;

▷ new Cologne–Frankfurt line with 40‰ maximum down gradient.

The design calculations were carried out for cast iron discs (with organic linings) and for cast alloy steel discs (with sinter metal linings). In comparison with the former, cast steel brake discs are able to absorb considerably higher brake forces or energies.

The characteristic advantages, as confirmed by test rig and practical trials with both the ICE/V and the TGV, are:

▷ higher temperature stability, less tendency to heat crack;

▷ less sensitivity to moisture;

▷ more uniform friction coefficients at high speeds; and

▷ relatively little wear of the brake lining at high brake forces (but higher wear at low lining load).

Thermal design calculations produced the result that:

▷ in the second batch of the ICE four cast steel brake discs with sinter metal linings are necessary per carrying wheelset;

Fig. 3: MD 530 carrying bogie with disc brake

▷ whereas in the first batch, three discs per carrying wheelset are adequate when cast steel with sinter metal linings. An alternative is a variant with four discs per axle in cast iron in conjunction with organic linings.

In order to reach a standardized structural concept for both ICE production batches, the four-disc arrangement per carrying axle was chosen. The first batch will still be fitted with cast iron discs; for the higher loads to which the second batch will be subjected, however, the switch to cast steel discs is unavoidable.

The construction drawing in Fig. 3 shows the ICE carrying bogie with a total of eight ventilated brake discs of 80 mm thickness and a single calipar per disc with an 8″ brake cylinder. The handbrake acts through a separate calipar independent of the pneumatically operated brake calipar on a brake disc on one axle in one of the two bogies on the vehicle.

Electro-magnetic track brake, linear eddy-current brake

Brakes acting on the wheels, such as the regenerative or the disc brake, depend for their achievable braking force on the available wheel/rail coefficient

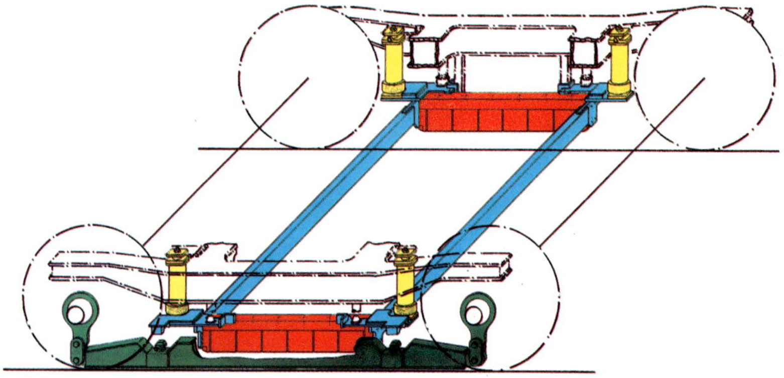

Fig. 4: Linear eddy-current brake in the trailer bogie

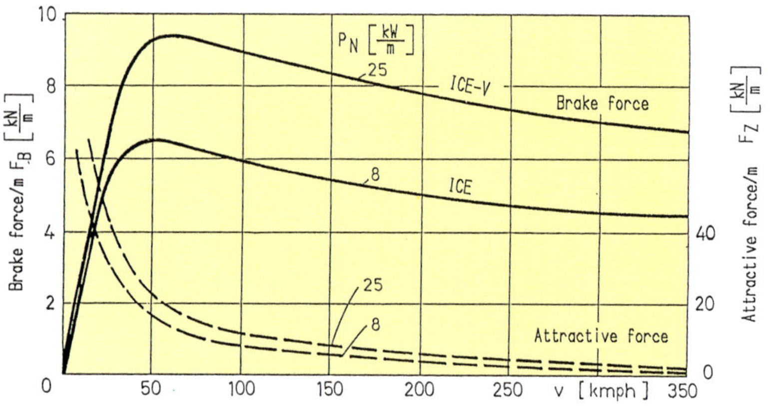

Fig. 5: Characteristic curves of the linear eddy-current brake
air gap μ = 7 mm, excitation output/m P_N at = 20°C

of adhesion which, as is well known, is influenced by the weather and reduces with increasing speed. To ensure that the ICE comes reliably to standstill within the braking distance during emergency braking, even in unfavourable weather conditions, an additional braking system, which does not depend on the wheel/rail coefficient of adhesion, is required.

Here, the electro-magnetic rail brake, used, for instance, in the IC trains of the Federal German Railway (DB), and the linear eddy-current brake, installed for the first time in the ICE/V, are available. Because the electro-magnetic rail brake is subject to considerable wear and is not adjustable, it must be applied only during emergency braking and cannot be used as a gradual brake. At the relatively low power requirement of about 1 kW per magnet the lead battery, which is standard equipment on the passenger coaches, can be used for the current supply.

The eddy-current brake, a new development of the Knorr Bremse AG, is a novelty in rail vehicles and was first tested by the Federal German Railway in the ICE/V as a service and emergency brake system.

Fig. 4 shows the suspension construction. Two eddy-current magnets each form a frame with track rods, which is suspended from mounting cylinders in the bogie.

In the braking mode, the frame is supported by traverses mounted between the axle bearings; the traverses constitute a constant level relative to the track. The support bearings on the traverse are vertically adjustable, so that the air gap, which changes with wheel wear, can be readjusted to the tolerance value of 7 ± 1 mm during general inspections.

It was possible to demonstrate the technical advantages of the eddy-current brake during several years of ICE/V trials:

▷ favourable brake force characteristic curve;

▷ independence of the coefficient of adhesion (in any weather conditions, during winter with snow and ice, always reliable, high braking efficiency);

▷ wearfree and continuously adjustable.

The considerably higher power consumption of the eddy-current brake compared with that of the electro-magnetic rail brake must be noted: it increases at the square of the brake force (Fig. 5, with comparison between ICE and ICE/V). The circuit concept planned for the ICE with eddy-current brake current regulation is derived from that of the ICE/V, with supply from the current bus bar via a phase-controlled converter with a series-connected transformer. Because such a circuit cannot guarantee the reliability of braking required with emergency braking – a central breakdown such as power failure would result in the failure of all eddy-current brakes – there is an automatic switch to the coach battery which operated with the aid of a diode circuit if the bus bar current supply is interrupted.

The first batch of the ICE production trains is still equipped with 110 v lead batteries. However with eddy-current brakes these have to be replaced by NiCd batteries, because the former are unable to supply the required electrical capacity at low temperatures and intense currents.

Other points to be considered are rail heating and effects on the signalling equipment [2].

The eddy currents generated during braking heat the rails and can affect the stability of the track. Permissible maximum temperature values, determined by the Federal German Railway – as a function of the type of rail and of the state of the rails – should, however, very rarely be reached on the new lines (cumulative effect of extremely strong solar radiation, short intervals between trains, braking always at the same spot).

In the procedure tested in the ICE/V, infrared sensors are mounted underneath the power units to measure the rail temperature. The measuring signal is evaluated by the brake control and, when the permissible temperature limit value is reached, it redistributes the brake forces, in that the disc brake takes over the eddy-current brake. Adequate temperature reserves are allowed for, so that for reasons of safety

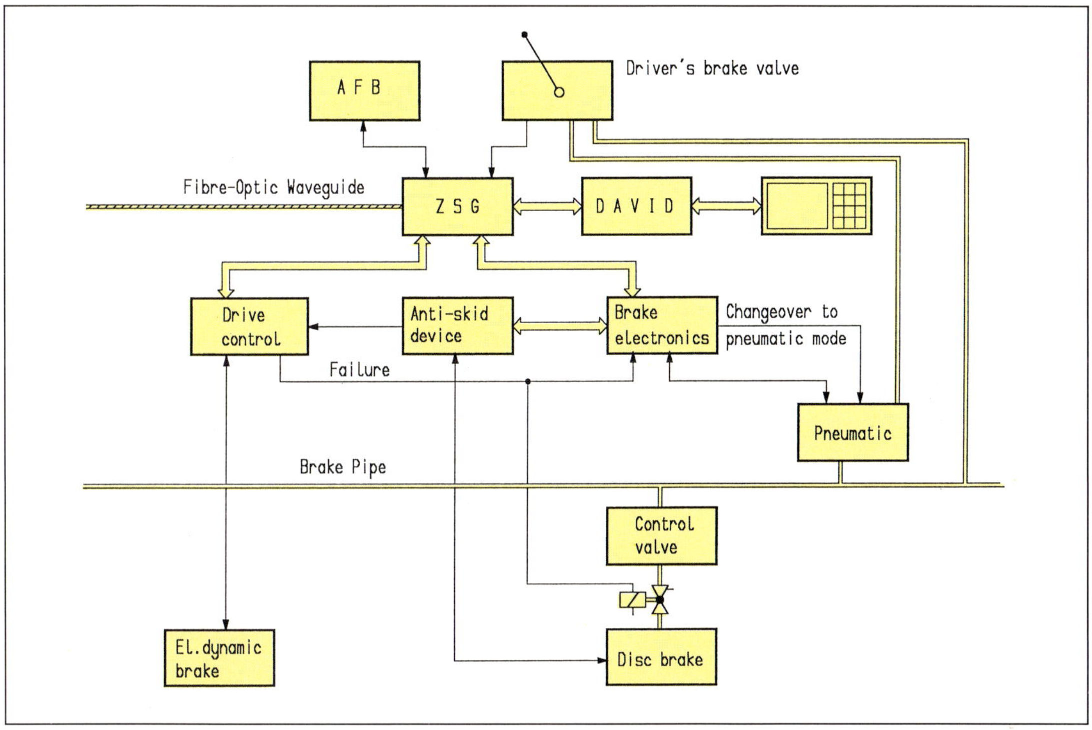

Fig. 6: ICE brake control, power unit

during emergency braking the eddy-current brake is still fully functioning.

Track investigations carried out by the Federal German Railway have shown that the electro-magnetic field of the eddy-current brake may interfere with the signalling installations (for instance with electric axle counters). Modern signalling equipment on the new lines is not affected by such interference. On the basis of the results of developments so far it can be expected that effects on the axle can be prevented by means of screening the eddy-current brakes. For this purpose, the latest tests carried out on the ICE/V are still awaiting evaluation. However, the problem of interference with signalling will become progressively less important because of the continuous modernization of the Federal German Railway signalling installations (which will be resistant to the eddy-current brake).

The decision on the eddy-current brake will be determined by cost effectiveness, which is most reliably demonstrated by line trials. The higher prime costs are compensated by lower maintenance costs owing to the elimination of the badly wearing electro-magnetic rail brake and through the considerably longer service life than that of the disc brake. The advantages of the eddy-current brake become evident especially on high-speed vehicles, so that in view of the current trend towards higher speeds, an interesting market will be created in this field. For the ICE the option of the eddy-current brake has been kept open; the trains can be converted for this brake without modification of the vehicle design.

In detail this means that the conditions for the installation of the current supply, for the suspension construction in the bogie, and for the brake control have all been allowed for. It is considered useful to equip some production trains with the eddy-current brake in order to demonstrate reliability and cost effectiveness in practical operation. A decision on this matter is still awaited.

Brake control

The brake control was developed and designed with the following decisive targets and functions in mind:

▷ high accuracy of brake force regulation and target braking;

▷ automatic speed control;

▷ a high level of comfort through jerk-free interaction between the various brakes;

▷ economical brake application on the priority principle;

▷ coupling of the brake control with the computer-assisted train diagnostic system;

▷ automatic brake test; and

▷ a high level of availability, uninterrupted operation even in the event of an electronics breakdown.

The ICE is equipped with an electronic brake control which meets these requirements. It is the Knorr microprocessor driver's brake system in conjunction with an automatic, electro-pneumatic pre-regulated compressed-air brake (type designation "HSM").

Structure

The brake electronics which is installed in a 19″ rack in the power unit (Fig. 6), is connected with the control desk, with the fibre-optic train data bus, with the central vehicle diagnostic computer (DAVID) and with the drive control (or the regenerative brake respectively) via the train control unit. A second RS 323 bus establishes the link with the separately arranged microprocessor anti-skid device – also in a 19″ rack – for the exchange of speed and of diagnostic information

In the centre trailer (Fig. 7) the 19″ electronic rack unit contains the anti-skid device, together with the nected in parallel to a three-phase busbar to which the two traction motors of a bogie are connected. In contrast, the power requirement of a bogie can be met by a single machine-side converter with the use of the high-capacity GTO converters.

Auxiliary equipment circuits

The main transformer is provided with an 885 v and a 200 v winding for the auxiliary equipment power supply.

Three 65 kVA static three-phase auxiliary converters, HBU, are connected to the floating 885 v winding to supply the fans and pump motors. On account of the stringent availability requirements, three autonomous three-phase power systems are provided which can be partially reconfigured (Fig. 4).

HBU 1 and 2 are allocated to the auxiliaries of bogies 1 and 2 respectively. Both inverters generate a voltage/frequency controlled three-phase power supply. Voltage and frequency are controlled in order to provide the necessary cooling output, thereby minimizing power consumption and fan noise.

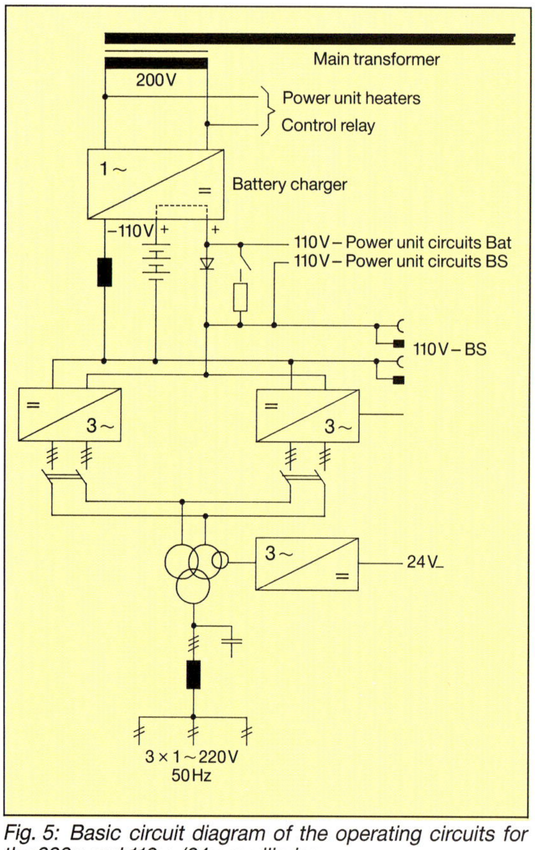

Fig. 5: Basic circuit diagram of the operating circuits for the 220 v and 110 v-/24 v-auxiliaries

The third converter, HBU 3, is operated at a constant frequency of 60 Hz and a voltage of 440 v; voltage/frequency control is employed in this case only for current limiting during starting after loads have been connected. This HBU supplies all auxiliary equipment requiring a power-dependent electrical supply or which do not tolerate speed variation as a result of the low thermal time constant of the equipment to be cooled. This class of equipment includes the line filter oil pump (NFÖP), both transformer oil pumps (TRÖP), the four filter fan motors (SKL) and the eight converter fan motors (SRL), both integral fans of the auxiliary inverters (HBUL), the charging reactor fan (LADL), the air compressor (LUPRE) and the compressor/evaporator unit (VDKL) and the evaporator fan (VVKL) for the driving cab air-conditioning plant.

Power supply to a bogie is switched off on failure of a converter. If HBU is affected, its loads are changed over to HBU 2, which is then operated at a fixed frequency (440 v, 60 Hz).

The 200 v auxiliary winding of the main transformer supplies power to the heater windings for front windows, floor, air-conditioning plant, sand boxes and valves, the control relay and, in particular, the battery charger (Fig. 5). The output voltage of the charger is available after smoothing for the 110 v circuits of the power unit partly directly and partly decoupled through a diode for supply of the 100 v battery busbar of the trainset.

Power supply from the busbar to essential loads is maintained by means of this decoupling on failure of the charger. Should a fault occur e.g. of the high-voltage equipment, or at the main transformers of the leading power unit, this can still function as a control unit in order to maintain interruption-free operation. Battery charging is effected in this case via the battery busbar from the center trailer, for which purpose the decoupling diode is short-circuited.

Two independent static auxiliary inverters, each rated at 5 kVA, are used for line voltage independent power supply of the various inverter voltage loads. The switched power supply employs a common output transformer which is provided together with a winding to supply 24 v DC loads via a rectifier bridge.

Single-phase loads are connected between two phases of the 3 × 220 v, 50 Hz, power system downstream of an LC filter.

Components

Main circuit-breaker

A vacuum circuit-breaker was used for the first time in a Federal German Railways' vehicle for the main circuit-breaker in a high-voltage rack in the machinery compartment.

This circuit-breaker (Fig. 6) makes effective technical and economical use of the characteristics of vacuum arc quenching technology, i.e.:

▷ high switching performance, i.e. product of breaking current and number of switching operations;

▷ long operational life;

▷ long mechanical life;

▷ low minimum breaking current and consequently low switching overvoltages.

Technical data

▷ rated voltage: 17.5 kV;

▷ rated frequency: 16⅔ Hz;

▷ rated short-circuit breaking current: 25 kA;

▷ rated short-circuit making current: 63 kA;

▷ operating current: 2000 A;

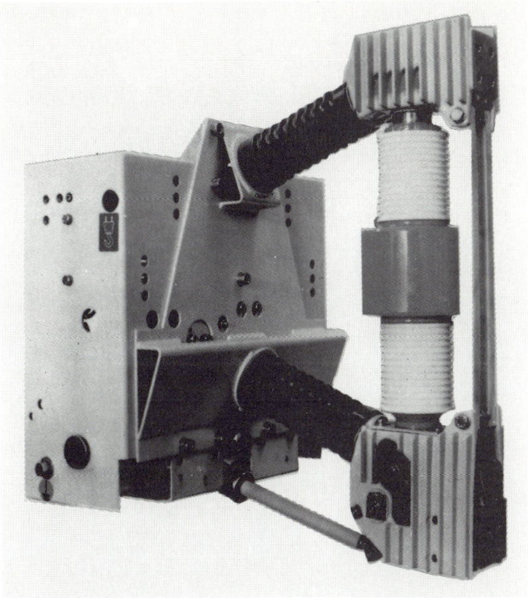

Fig. 6: Single-pole vacuum circuit-breaker 3AF9244

▷ arcing time: 3–33 ms;

▷ drive type: spring-loaded operator with motor charger.

Transformer (Fig. 7)

The power rating of the transformer was measured on the basis of simulated track runs specified by the Federal Railway. The power demand for the complete new track system for a train weight of 880 t resulted in an effective value of 4500 kVA. The transformer is loaded with a traction power demand of 5200 kVA in the difficult track section in the foothills of the German central mountain range between Kassel and Fulda. The driving time in this section lies in the vicinity of the thermal time constant of the oil and represents a multiple of the thermal time constant of the winding. These factors had to be taken into consideration in determining the size of the transformer.

In the type test it was proved that the winding temperature rise under converter operation conditions lies considerably below the permissible limit of value for heat-proof insulating material.

The transformer is of core-type design similar to that for the IC Experimental. All the design features of the lower rated previous transformer, which has adequately proved its value in operation, were maintained.

A sudden short-circuit test was performed within the scope of the extensive type tests. It was thus possible to prove that no deformation of the windings or of the leads occurred as a result of the magnetic forces involved.

The major technical data are collated in the following table:

	Unit	Primary winding	Traction winding	Heating winding	HB 1 winding	HB 2 winding
rated voltage	V	15000	4 × 1432	1011	884	200
rated current	I	348	4 × 787	495	199	180
rated power	kVA	5220	4 × 1127	500	176	36
Total weight including filter reactors and oil filling: 10 t						

GTO thyristor converter

Converter technology employing voltage-source dc-link circuits experienced a major innovation with the introduction of gate turn off (GTO) thyristors. The largest element nowadays employed (Fig. 8) characterized by a blocking voltage of 4500 v and a breaking current of 3000 A allows implementation of an output of approx. 2700 kW in a three-phase bridge-circuit of a single inverter unit. In this circuit neither series nor parallel connection of semiconductors is necessary.

Fig. 7: Transformer

The transition from thyristor-controlled to GTO thyristor technology was made easier by the fact that the DC-link-voltage of 2800 v, chosen for the thyristor-controlled converter, could be retained. As a result, it

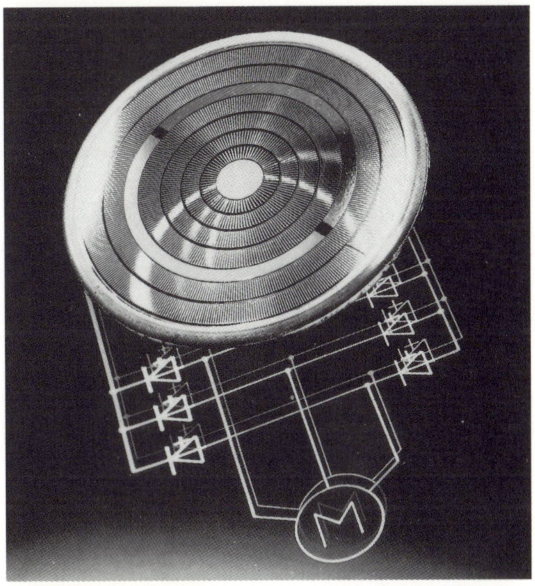

Fig. 8: Silicon wafer of a GTO 4,500 v/3,500 A

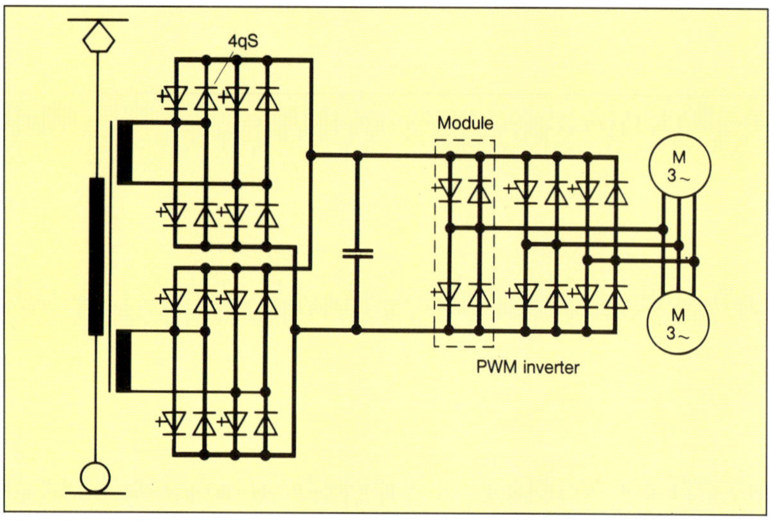

Fig. 9: Block diagram GTO voltage source inverter (application) ICE

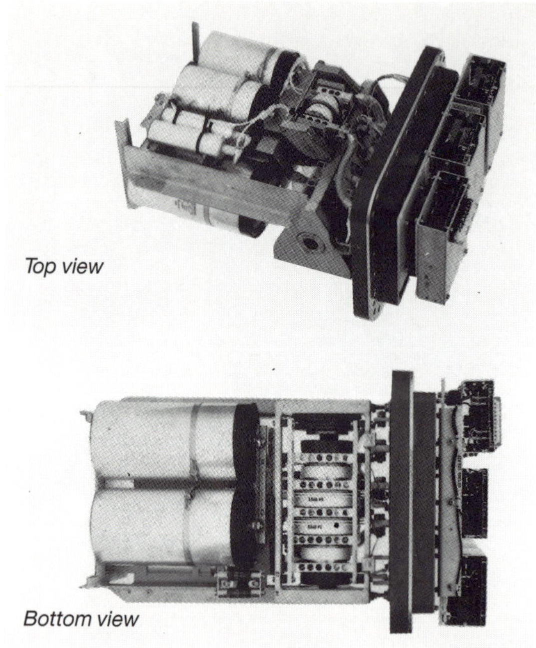

Fig. 11: Converter module with GTOs and evaporation-bath cooling (withdrawable active part)

was not necessary to modify the other components of the traction plant, such as line-side filter, transformer, DC-link and traction motor.

The complete GTO converter plant for a bogie comprises two 4-quadrant power controllers employing time staggered triggering and a pulse width modulated inverter; only seven phase modules are required for the complete configurations (Fig. 9).

The phase module employing GTO thyristor technology in the power circuit is of simple design. This can be seen in comparison with a thyristor module (Fig. 10). It must further be taken into account that four elements connected in series were required with a thyristor unit.

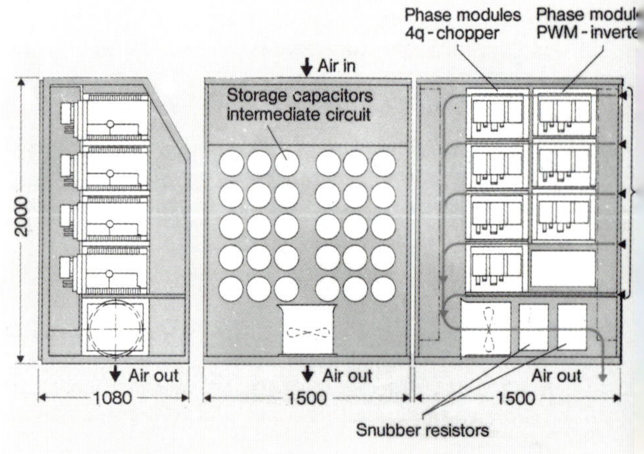

Fig. 12: Converter (4q-chopper + PWM-inverter) with GTO-thyristors and evaporation bath cooling for one bogie of ICE

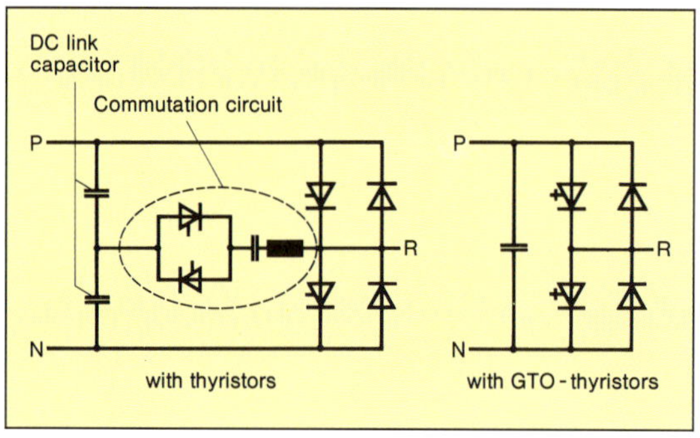

Fig. 10: Comparison of one inverter phase

Fig. 13: GTO converter for one bogie of ICE

Fig. 11 shows the converter module. The four semi-conductor elements can be clearly seen; they have been integrated in a column together with the evaporation coolers.

Evaporation bath cooling is employed in the ICE for removal of the heat losses as for other high-power converters. This cooling method is a particular advantage where considerable heat losses occur in a confined space.

Fig. 12 depicts the layout of the phase modules and of the DC-link capacitors, including the associated air ducts.

Significant technical improvements were achieved in addition to economical advantages with the adoption of GTO thyristor technology:

▷ the total weight of a power unit was reduced by 3.5 t for the same output;

▷ efficiency was improved, since the conversion losses were reduced to approximately one third;

▷ the controls were simplified;

▷ the number of semi-conductor elements was drastically reduced.

In order to ensure trouble-free application of this technology, the complete converter plant was subjected to extremely severe power tests under actual operating conditions.

The test set-up of the converter plant, including the filter reactor frame, is shown in Fig. 13.

In a second step, two power cars (numbers 401 051 and 401 551) were equipped with GTO thyristor converters, even before the start of production in order to obtain actual operating experience.

Traction motor and motor series reactor

The design of the traction motor is based on a motor employed in the experimental trainset. The appreciably larger power demand of 1250 kW, compared to 700 kW per motor in continuous operation, necessitated optimum utilization and cooling of the motor, due to the fact that the unit volume and weight could not be appreciably increased.

The motor has the following data required to satisfy the requirements of the traction force diagram of Fig. 1:

		Continous	V_{max}
Type		BAZ 7096/4	
Rating	kW	1250	1146
Voltage	v	2050	2200
Current	A	415	318
Speed	rev/min	1535	4032
Torque	Nm	7775	2713
Power factor ..		0.88	0.82
Weight	kg	2010	

The traction motor is designed as a 4-pole induction motor with forced ventilation. The stator (Fig. 14) is of pot-type welded construction into which the stator laminations were inserted by means of a press. Ventilation of the stator core is effected from the non-drive side via channels at the circumference of the laminations and through the air gap.

Fig. 14: The stator of the ICE-traction motor

Fig. 15: The rotor of the ICE-traction motor

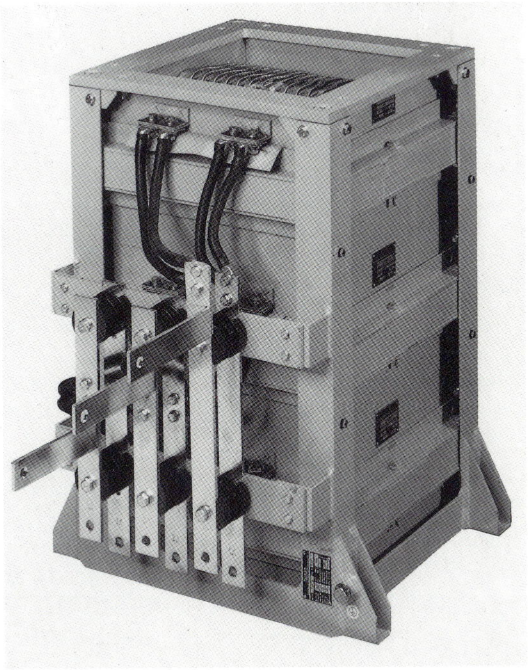

Fig. 16: Motor series reactor

Temperature detectors are installed in three slot keys for monitoring of the motor temperature. The motor connections and the measuring leads are routed to the machinery compartment through the air inlet cowl.

The rotor (Fig. 15) is designed as a squirrel cage rotor with copper bar of trapezoidal cross-section and is ventilated by means of two rows of holes in the laminated core assembly.

The rotor speed is detected by means of a speed sensor and the temperature of both traction motor bearings measured by means of Pt 100 RTDs. The non-drive side bearing is grease lubricated and the drive-side bearing is oil lubricated in conjunction with the gearing.

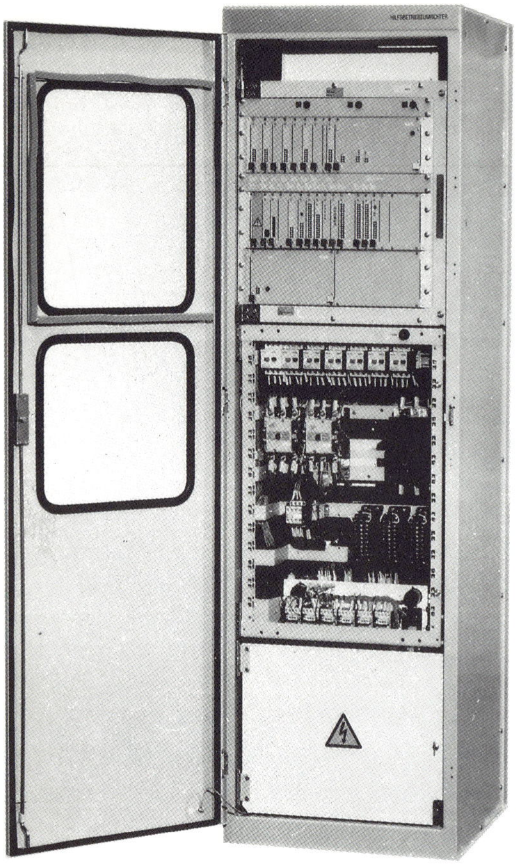

Fig. 17: Auxiliaries converter unit, 65 kVA

A three-phase motor series reactor (MVD) (Fig. 16) is connected in series with the motor up to a motor frequency of 70 Hz. It is force ventilated parallel to the traction motor: the air requirement of the traction motor is 1.0 m³/s. The reactor comprises three individual reactors connected in series on the air side. The total weight of the motor reactor is only 200 kg for 3 × 0.92 mH, 400 A (cont.), as a result of radical weight reduction measures.

The design employed allows continuous operation of the traction motor on the hyperbolic power characteristic at 1250 kW in the speed range above 110 kmph; the motor series reactor is short-circuited at speeds in excess of approx. 150 kmph.

Auxiliary equipment converter

The three-phase power systems with both variable and constant voltage and frequency required to supply the three-phase induction motors of auxiliary equipment is taken from three nearly identical auxiliary equipment converters (HBU) employing a type tried-and-tested in operation on the E 120 locomotives. Each HBU (Fig. 17) is capable of continuously supplying 65 kVA at a power factor of 0.8. It was thus possible to employ individual modules identical to those employed in the E 120.

The control electronics, which were also retained, were supplemented by an interface card for connection to the train diagnostic equipment. The controls and the power sections are installed in a cabinet (600 × 600 × 2000 mm, 370 kg) together with a swing-out frame, which contains the three-phase distribution. The three-phase distribution is specific to the cabinet due to differences in the load connected to it.

The dc-link reactor is installed behind the cabinet. A common ventilator cools the HBU and the reactor, using fresh air. Cooling is monitored by means of temperature sensors, as for the E 120. In addition, air-flow monitoring is provided in conjunction with sealing of the exhaust channel when the power unit is driven through the wash-plant.

Battery charger

An IUU characteristic was specified for high rate charging and trickle charging of the train battery type 2 PzS110. The battery charger type employed in the E 120 series was adapted with no change of dimensions (Fig. 18). The current stage charges uses a variable charger current limit with superimposed battery current limiting, based on the five hour discharge current (I_5). The first voltage stage operates at a constant charging voltage of 2.35 v/cell until the charging current has fallen to 0.5 × I_5. Thereafter, the second voltage stage operates at a trickle charging voltage on 2.23 v/cell. If the cell voltage drops to 2.1 v, automatic changeover to high rate charging at 2.35 v/cell is effected. The voltage values are compensated by means of a temperature sensor on the battery. If the battery temperature exceeds 55° C during high rate charging, automatic changeback to trickle charging is effected and this is signalled to the diagnostic equipment.

Fig. 18: Battery charger

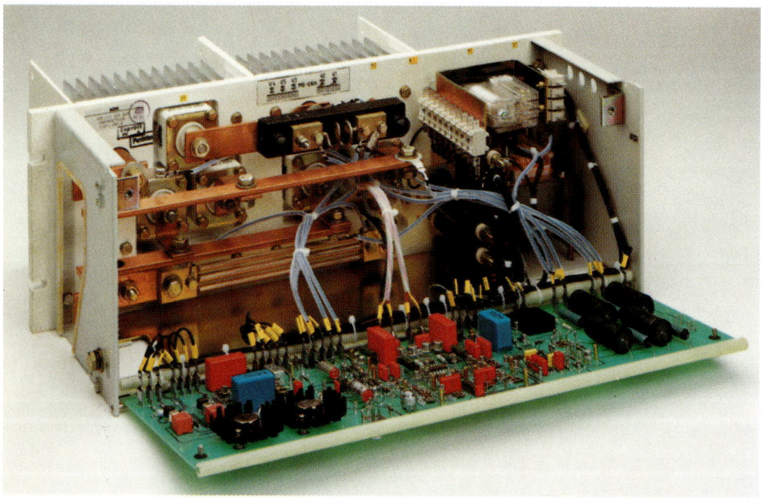

Auxiliary inverter (HWR)

The two redundant static auxiliary inverters (Fig. 19) generate an independent three-phase power supply of constant voltage and frequency from the variable train battery voltage.

Technical data	Type HWR 110/2,3	
Input voltage	110 v	+40 % −30 %
Output voltage	3 × 220 v_{rms}	± 5 %
Frequency	50 Hz	± 2 Hz
Rating	5 kVA	
Weight	45 kg	

In order to meet the power requirement of the power unit, two transistorized power sections, each rated at 2.5 kVA, are controlled in parallel and are interconnected via the output transformer. The power sections are of identical design to those for the auxiliary inverters in the intermediate coaches.

The mechanical design employs a 19″ packaging system with a tier height of 445 mm (10 height units as per DIN 41 494) and 350 mm deep. An INTERMAS-

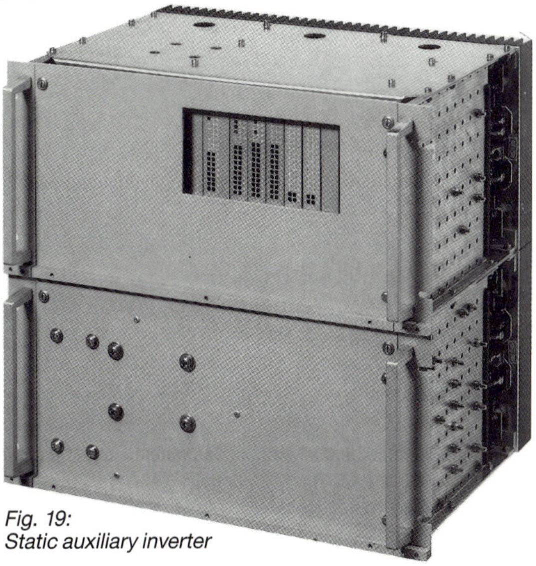

Fig. 19:
Static auxiliary inverter

cassette with six type C control plug-in cards is integrated in the card sub-rack.

The output transformer and filter are separately mounted in the vicinity of the auxiliary inverter, as are the three-phased contactors required for switching the redundant equipment.

Instrumentation and control equipment

Train control unit (ZSG) and data transmission

Tasks

The primary task of the train control unit (ZSG) is the acquisition and transmission of data required for control and diagnostics between the instrumentation and control equipment.

These are (see Fig. 20):

at the train coordination level:

▷ the driver's console;

▷ data of the automatic train control system (LZB) indirectly via the DAVID system (see below);

at the train control level:

▷ the automatic motoring/braking controls (AFB);

▷ the central position and speed detection (ZWG);

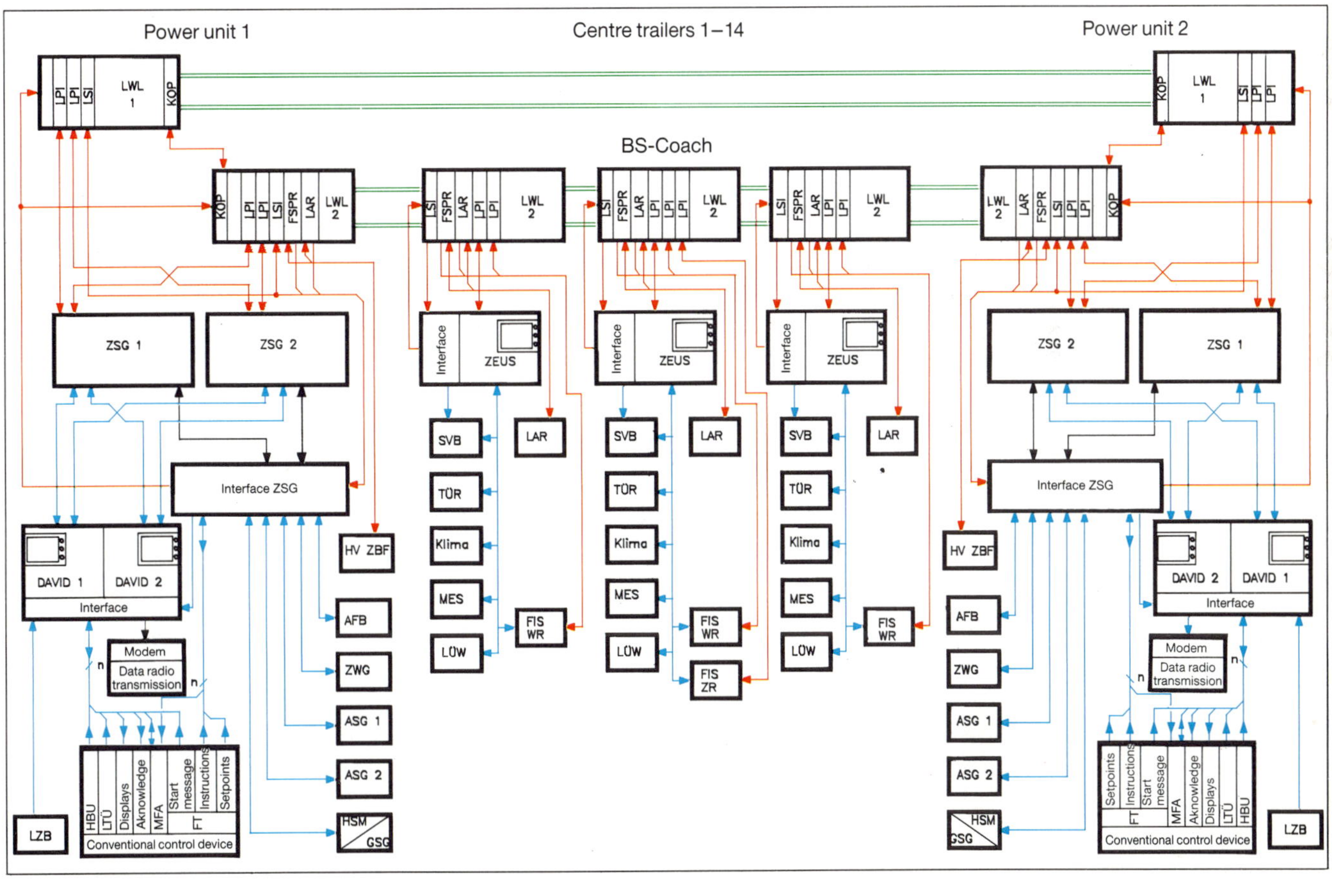

Fig. 20: Guidance technology, IC Express

Legend:

ASG	Drive control
ZSG	Train control
LZB	Continuous automatic train control
HVZBF	Main distributor ground train radio
HSM	Type designation for driver's brake valve appliance
GSG	Wheel slide protection
MFA	Modular driver's cab display system
HBU	Auxiliaries converter
LTÜ	Bearing temperature monitor
ZWG	Central distance and speed indicator
FT	Driver's console
AFB	Automatic traction- and brake control
LWL	Fibre optics station
LPI	Fibre optics processor interface
LSI	Fibre optics signal interface
FSPR	Interface telephone
KOP	LWL coupler
LAR	Interface loudspeaker installations in the passenger coach
DAVID	Diagnostic and preparation service with integrated display control
ZEUS	Central installation for monitoring and control
SVB	Signal processing brake
FIS	Passenger information system (audio/video system, telephone with C-network)
MES	Measured-value indication system
LÜW	Runnung state monitor
KLIMA	air-conditioning plant
TÜR	door control unit

▷ the equipment for diagnostics, putting vehicle into service, programme preparation with integrated display (DAVID);

at the vehicle control level:

▷ the two traction control units (ASG);

▷ the electronic driver's brake valve plant (type designations HSM);

▷ connected via the trainbus (employing opto-fibre conductors) with the control equipment of the non-leading power unit and to the central acquisition, monitoring and control equipment (ZEUS) in the intermediate coaches.

A further important task for which data acquisition is required is logic checking of status signals, i.e.

▷ of interlocks;

▷ of the running/braking logic circuits;

▷ and detection of inconsistencies.

The last point in particular is of importance for self-diagnostics by micro-processors, in order to detect fault conditions in the equipment and, where this can be achieved with reasonable investment, in peripheral equipment and at the equipment interfaces. This task not only involves transmission of alarms to the diagnostic computer system (DAVID), but also transmission of the status of the transmission paths to all equipment connected to the bus, so that effective countermeasures can be taken by the equipment in the event of a data transmission fault.

Structuring

A requirement in the structuring of the instrumentation and control equipment was that a single fault at any point in the system should not result in complete failure of the train controls.

The train control unit is consequently of fully redundant design (1 out of 2 structure); changeover from one unit to the other is effected manually by the driver, who is assisted in his choice by self-diagnostics and by the DAVID diagnostic system. The opto-conductor train bus was made partially redundant and has been configured as a ring on the basis of cost/availability optimization and uses automatic changeover on occurrence of a fault. The power unit interfacing equipment and the transmission paths are redundant; the intermediate coaches' equipment is non-redundant. In the event of a single fault, the sole consequence may be loss of the remote-control capability of an intermediate coach, which is of no consequence for operation of the trainset.

Interfaces

The train control unit has a large number of parallel and serial interfaces, which were selected on the basis of data quantity and data type, in view of its central role in the instrumentation and control system.

The following interfaces are provided per train control unit:

for the driver's console and for the hard-wired level:

▷ 56 conventional 110 v inputs;

▷ 40 conventional 110 v inputs;

▷ 2 analogue setpoint inputs with monitoring;

▷ 2 analogue actual value outputs;

to DAVID:

▷ 2 serial high-speed interfaces with a gross baud rate of 500 kbit/s as EIA RS 485 and BN 411 004 part 4;

to the opto-fibre train bus:

▷ 2 high-speed serial interfaces, 500 kbit/s as for DAVID;

to the automatic motoring/braking controls (AFB), train control unit (ZWG), 2 traction control units (ASG), and the driver's brake valve plant (HSM):

▷ 5 serial current loop interfaces with a gross baud rate of 4.8 kbit/s, conforming to the industry 20 mA TTY interface standard.

On changeover of the train control units, the outputs of the conventional interfaces and of the 20 mA interfaces are changed over, while RS 485 interfaces and all inputs are left permanently connected.

The nodes of the opto-conductor train bus employ a parallel computer bus as a system interface to which special interface cards can be connected. Each opto-conductor train bus node provides facilities for ten interfaces and the total gross baud rate of the train bus is 5.12 Mbit/s.

The power units are fitted with the following standard equipment:

▷ 2 opto-conductor processor interfaces (LPI) for data transmission via RS 485 interface at 500 kbit/s to the train control units;

▷ 1 opto-conductor signal interface (LSI) with a parallel conventional eight bit wide input/output interface, for operation of the electropneumatic brakes and of the train control changeover circuit in TK 2;

▷ 1 audio frequency interface telephone (FSpr) for connection of a private service telephone;

▷ 1 audio-frequency interface for the public address system in passenger coaches (LAR).

Only one each of the last two interfaces are provided in each power unit.

Data organization

Data transmission is effected on a cyclic basis. The predetermined cycle is 100 ms for all data processed by computers, and 1 ms for transmission on the opto-fibre conductor train bus.

Two different data types are employed within a data block:

▷ function data, whose significance is based in their relative position within the data block and which are periodically refreshed;

▷ diagnostic data, which are coded employing an agreed fault code and which are transmitted only on occurrence or disappearance of a fault.

Organization of the data into two types offers two advantages; namely, permitting the necessary cyclic refreshing of data required for running/braking functions; and utilization of existing data transmission paths for transmission of diagnostic information.

Data content

A description of all data communication involved would be outside the scope of this chapter. Data communication between the train control unit and a traction control unit in the leading power unit can be taken as an example of such data communication (see Fig. 21).

The transmission mode between the train control unit and the traction control unit will first be considered (left-hand column):

The train control unit transmits clock time in the first two bytes A1 and A2. The clock time is provided by the DAVID diagnostic system and is primarily intended for diagnostic purposes.

Bytes A3 and A4 contain the actual speed obtained from the train control unit (ZWG).

Bytes A5 and A6 contain the set-point for the drive. Depending on the polarity sign, this set-point designates traction force or braking force and is obtained from different sources, depending on the operational conditions. On operation without automatic running and braking, the set-point is taken from the set-point potentiometer on the driver's console, which is acquired by the train control unit and which is passed on in normalized form to the traction control unit on completion of checking functions which encompass wire break. On braking, the braking force value is obtained from the driver's brake valve unit, which receives the overall set-point from the train control unit, in a similar manner to that described above for the running set-point, and distributes the individual set-points to the various brakes. The driver's brake valve unit obtains the braking regenerative power from the traction control unit and passes this value on to the train control unit.

In operation under automatic running/braking control, the running and braking set-points originate in this unit and are calculated on the basis of the speed set-point and on information derived from the ATC system. This unit also issues set-points to the various types of brakes. However, manual operation of the braking lever has priority and immediately re-activates the circuit, i.e. braking lever – train control unit – driver's brake valve unit – train control unit – traction control unit.

The next byte A7 of the data block contains commands issued at the driver's console, each of which is coded employing two complementary bits to ensure data integrity, so that transmission errors immediately lead to impermissible bit combinations, thus ensuring reliable initiation of safety measures.

Bits 0 to 2 of data byte A8, comprise information from the ATC system, which describes the track quality (old section, upgraded section or new section) and thus the permissible contact wire current (600, 800 and 1000 A, respectively). The traction control unit ensures that the relevant current limit is not exceeded. Bits 6 and 7 contain complementary status information as to whether the driving cab is occupied and open.

The last byte (A9) is the so-called „topology byte". It provides information as to whether the relevant devices have been switched on and are functioning correctly as regards data communication. This byte confirms validity of the data transmitted.

In the receiving mode from the traction control unit and the train control unit (right-hand column), the diagnostic information already mentioned is contained in bytes A1 and A2, i.e. the code number of a new or reset fault of the traction control unit together with some additional information.

Then follows the internal set-point (A3 and A4). The traction control unit calculates the maximum possible set-point on the basis of the actual speed and on the traction force or braking force characteristic and relates this value to the value received from the driver's console. The resultant value is transmitted to the DAVID diagnostic system by the train control unit. The relevant actual value is coded in bytes 5 and 6; the train control unit adds the values of the four traction control units of the train set separately for

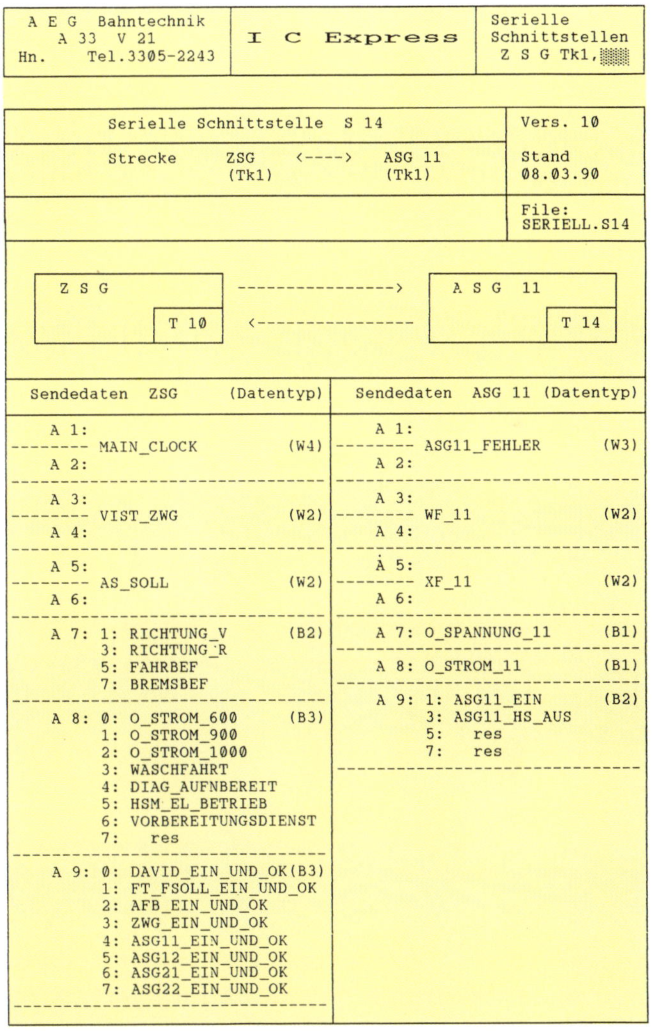

Fig. 21: ICExpress, serial interfaces

each power unit and also for the train set and transmits these values to the DAVID diagnostic system and indicates these values in the instrument on the driver's console after conversion into analogue form.

Bytes A7 and A8 contain the digital values of contact wire voltage and current measured by the traction control unit, which are passed on to the DAVID diagnostic system via the train control unit.

The traction control units sends its "heart-beat signal" in bits 0 and 1 of byte A9 to the train control unit, which integrates this in the "topology byte" and passes it on to all major items of equipment.

By means of bits 2 and 3 the traction control unit can trip the main circuit-breaker under defined conditions.

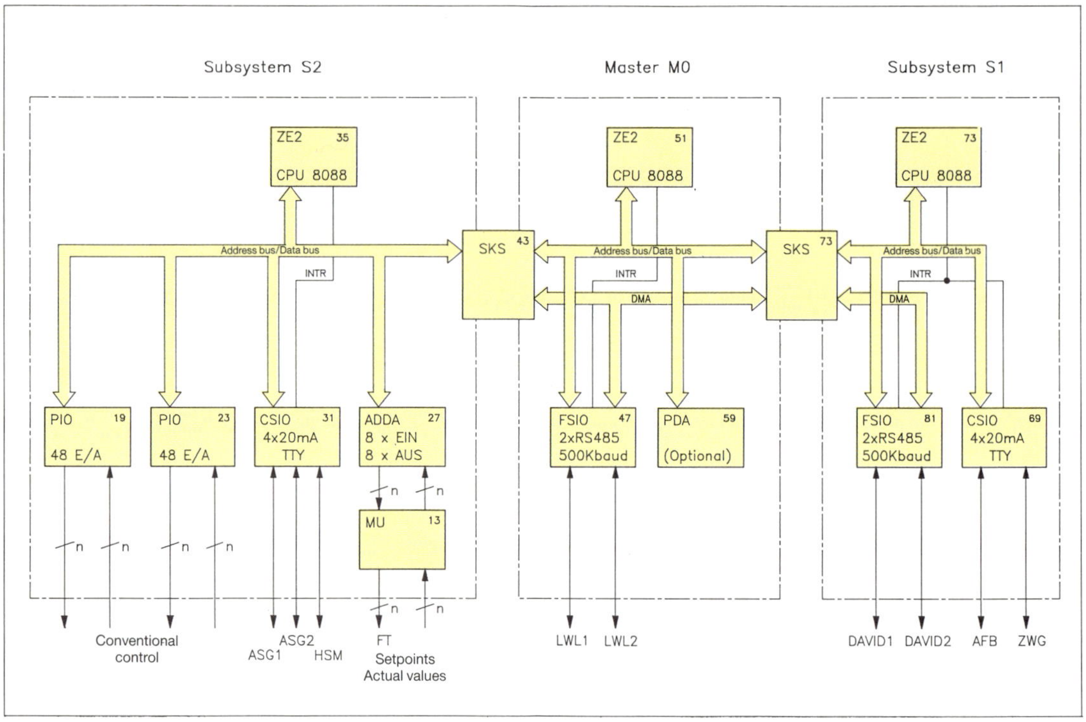

Fig. 22: Guidance technology ICExpress

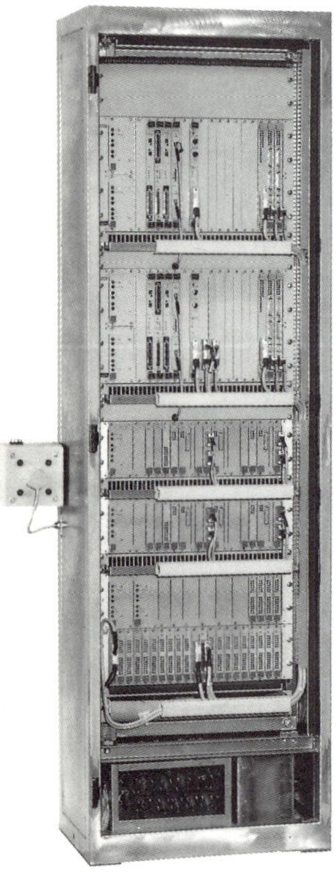

Fig. 23: Train control unit with fibre optic cable

Each data block is additionally provided with a start byte for synchronization of data transmission and with a check-sum to monitor data integrity.

Train control unit computer structure
(Fig. 22)

The train control unit employs plug-in cards of the GEATRAC system, which was developed on the basis of the CC 85 equipment employed in the IC Experimental.

A system comprising 3 computers linked via memory is employed for data processing and for servicing the numerous interfaces.

The master system (MO) performs the task of data processing and servicing the train bus. Provision has been made for installation of a process data recorder (PDA). The subsystems S1 and S2 mainly perform controller functions for the interfaces to the trainset. Subsystem 1 controls the series interfaces to the equipment of the train control level. Subsystem 2 services the equipment of the vehicle control level and the conventional interfaces to the 110 v controls and to the driver's console.

The measured value conditioning card is employed for matching the analogue inputs/outputs to the driver's console; this card also measures monitored criteria of the driver's console, the heat exchanger fan and temperature.

Installation of the train control unit and opto-coupler cards in the electronic cabinet

The train control equipment and the opto-coupler data transmission links are installed in a completely enclosed HF-proof cabinet (degree of protection IP 54) provided with its own cooling circuit and heat exchanger (Fig. 23).

HF-proof cabinet design

To provide protection against unavoidable EMI in the machinery compartment, the electronic cabinet is provided with metallic seals which ensure damping of at least 40 dB in the frequency range between 1 und 3 MHz.

All joints are provided with overlapping toothed strips and the door joints are provided with sealing strips of wire mesh.

All leads to the cabinet, with the exception of some short connections to the neighbouring cabinet, are routed through the cabinet wall by means of filter-type penetrations. Ferrite inductances and discharge capacitors are employed as filters in these penetrations and are effective in the frequency range described above.

The wire connections to the DAVID computers in the neighbouring cabinet are not filtered; data transmission at 500 kBaud lies in a frequency range in which filtering would lead to adverse effects. However, these leads are so short that only relatively weak interference is to be expected.

The leads to the ventilator monitors are also unfiltered.

The ventilation concept

The interior of the cabinet is sealed when the door is closed: it is provided with an internal forced air circulation system. The heat losses are extracted by a heat exchanger installed in the rear wall and which is externally ventilated, thus passing the energy to the atmosphere.

Fig. 24 depicts a cross-section with air flow.

The heat exchanger separates the cabinet interior from the atmosphere. It comprises a system of square cross-section tubes which are alternately traversed by cabinet and atmospheric air. The ventilators are mounted at the air inlet of the pipe system.

The air channels can be seen in Fig. 24:

The cooled cabinet air leaves the heat exchanger at A and enters the equipment compartment of the cabinet beneath the card subracks (BGT). Some of the air flows through the lower card subrack and part through a channel formed by the front panels of the card on the one hand and on the other hand by the front wall of the cabinet with door. Some of this air flows through the upper card subracks.

Air baffles are provided between the card subracks which deflect the air flowing upward from a card subrack around the next card subrack and into the space formed by the rear side of the card subrack and the inner wall of the heat exchanger.

This arrangement ensures that air which has been warmed in one card subrack does not flow through the next higher card subrack so that each card subrack is supplied with freshly cooled air from the heat exchanger.

The recombined air-flow from the main air duct reaches the inlet of the heat exchanger at B, where it is drawn in by the ventilator and forced into the heat exchanger duct.

On the external air-side the air is drawn in from the machine compartment at C and is blown into the external heat exchanger circuit. The warmed air exits once more at D.

As can be seen from Fig. 24, the heat exchanger employs counterflow between internal and external air-circuits, as a result of which the temperature difference is practically constant at all points, ensuring the largest possible heat transfer.

Fig. 24: ICExpress guidance technology, scheme of cooling air ducting in the electronics cubicle

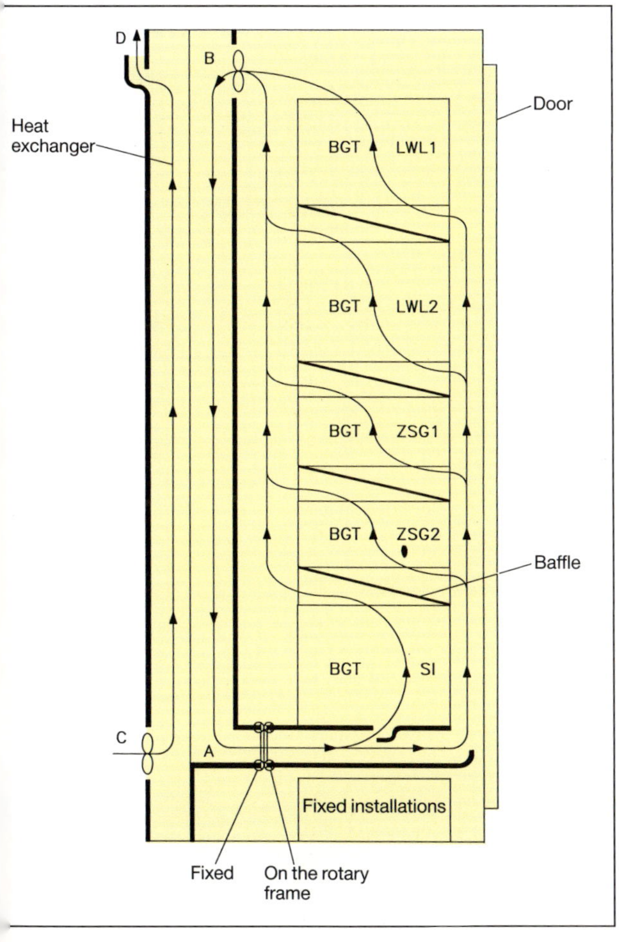

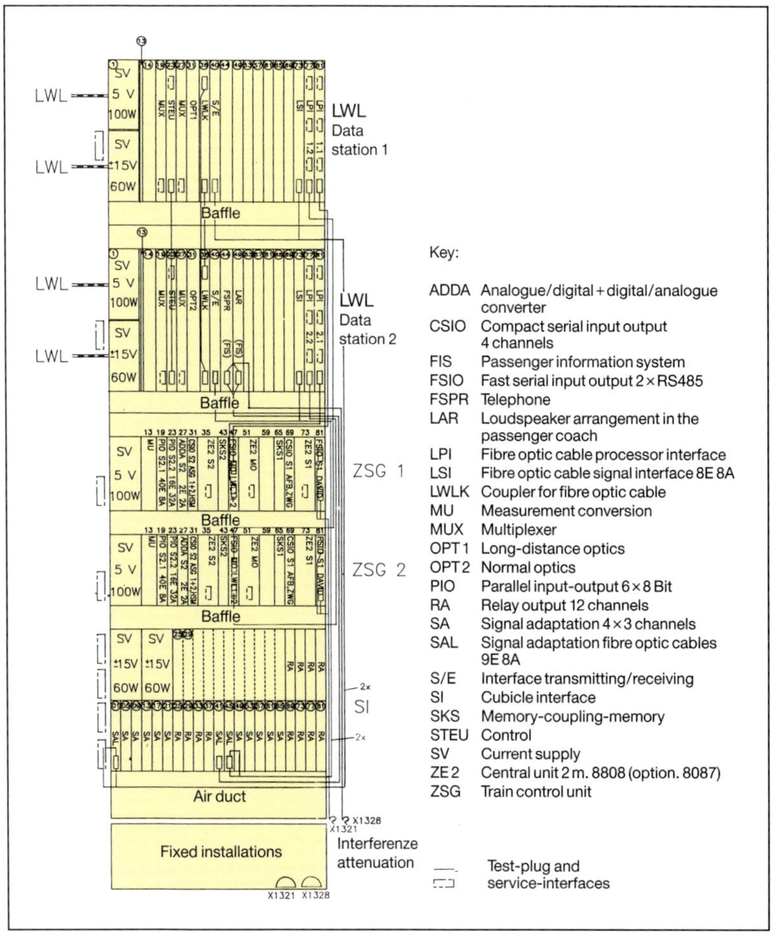

Fig. 25: ICExpress guidance technology

This concept does not require the use of decentralized card subrack fans and utilizes only central fans at points B and C. Two fans arranged side by side are provided at each of these two points. On failure of a single fan, the residual cooling power is sufficient to maintain operation, in the absence of excessive external temperatures.

Lay-out of electronic cards in the swing-out frame

The arrangement of the cards in the cabinet is depicted in Fig. 25 (front view).

This layout has been chosen with a view to obtaining the most favourable cable routes within the cabinet, so as to minimize EMI.

Since all cable penetrations of the cabinet are mounted completely below the swing-out frame, the signal interface cards, which are unique in that they have 110 v inputs and outputs in each plug connector, are arranged at the lowest level. As a result, optimum separation of the 110 v area from the signal processing area in the upper part of the cabinet has been achieved.

In the other card subracks, 110 v is required only for power supply to the power supply units. As a result of all the power supply modules being arranged at the extreme left-hand position, the 110 v leads are adequately separated from the plug-in cards.

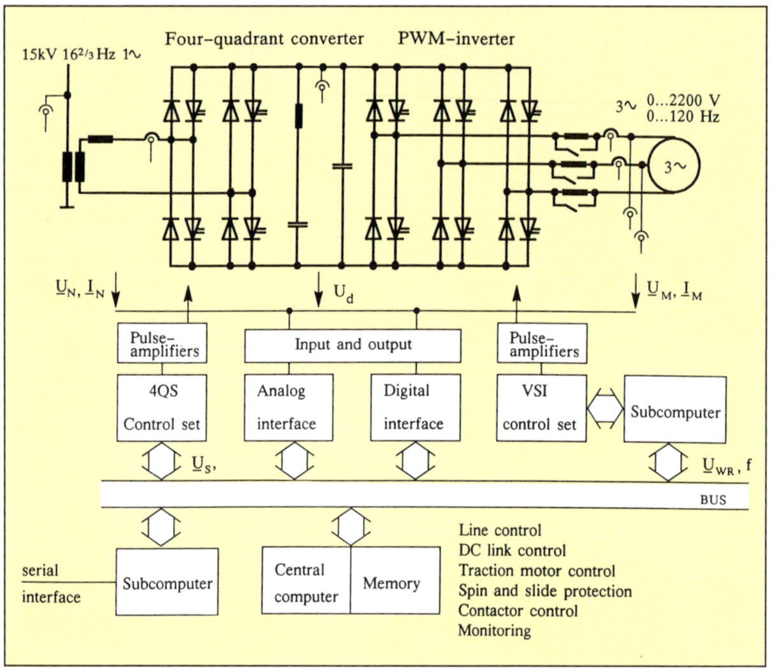

Fig. 26: Block diagram of traction control hardware for VSI

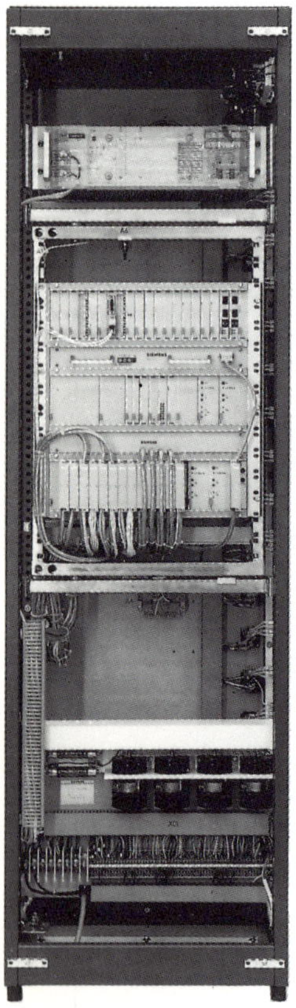

Fig. 27: ICExpress, SIBAS 1,6 MC traction control system

The other plug-in cards have been allocated on the basis of signal flow. The two train control unit card subracks are situated immediately above the signal interface, to which they are connected by a large number of cores. The opto-coupler units are installed at the highest level, together with the interface modules LSI and LPI, which are mainly connected to the train control unit.

Traction control unit (ASG)

The single-line diagram of the converter plant, together with the relevant controls, are shown in figure 26. The 4-quadrant power controller (line-side converter) is used to transfer active power from and to the supply power system (power factor = 1 in running and braking mode).

As a result of special clocking measures, the fundamental frequency of the supply current is in phase with the line voltage in the running mode and in opposition to the line voltage in the braking mode.

Since the pulse frequency is limited, the four power controllers employed in a power unit are clocked using time staggering, in order to reduce interference to the power supply system and to shift it into a higher frequency range.

The power made available by the power supply system is pulsed at twice the line frequency. The DC-link requires a constant power with respect to time for operation of the inverter. In order to reduce the size of the necessary energy storage unit, a filter circuit tuned to 33 Hz is used. The pulse width modulated inverter (motor converter) provides the induction machines with a three-phase supply of varying frequency and amplitude.

The primary task of the traction control unit is to act on the line-side and machine-side converters of the inverter plant in such a manner that the prescribed traction force and braking force values are obtained.

Design of the micro-computer train traction unit (Fig. 27)

The microcomputer unit employs elements of the SIBAS 16 system, which has been used with success for railway applications. The central processing unit and memory form the core of the system and coordinate all functions performed by the system. The microprocessor is responsible for line-side control, dc-link control, machine control, slide and slip control, contactor operation, monitoring and output of commands to the control units.

Previously, only one computer was employed for

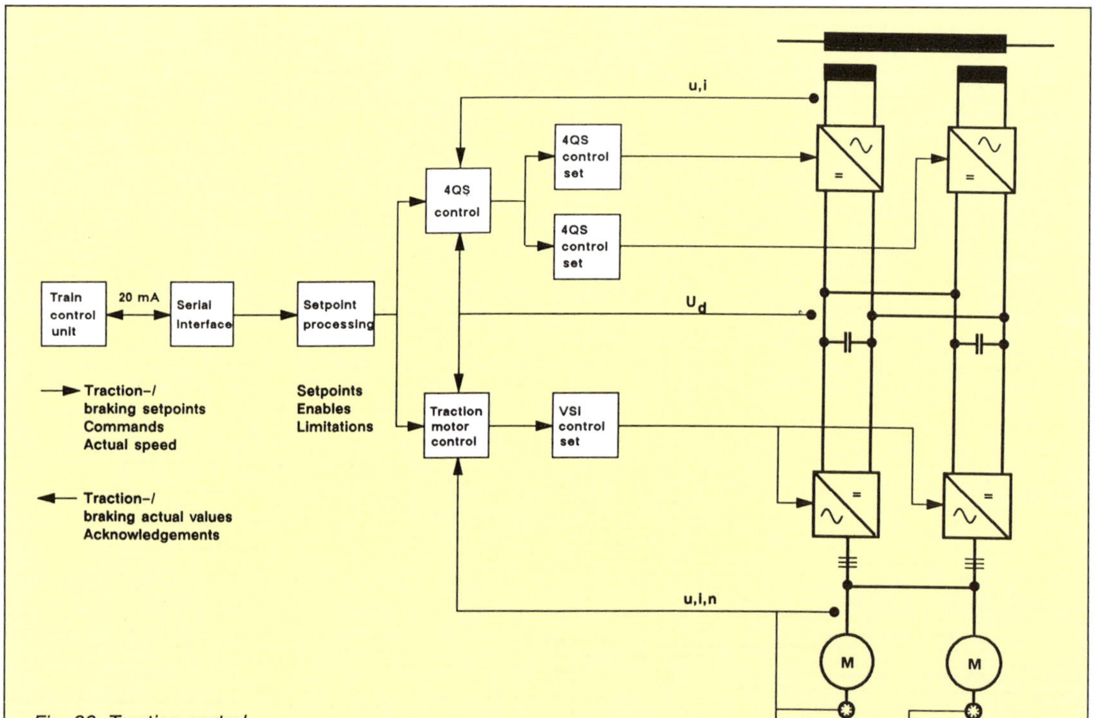

Fig. 28: Traction control

control purposes. In the case of the ICE controls, two sub-computers are employed which are coordinated by the central computer, and which service a serial interface to the train control unit and the pulse width modulated inverter control unit.

The link between the central area and the next level, i.e. the interface cards, is effected by means of a special system bus.

The pulse amplifiers (AIV) and input and output cards are located at the third level; these interface to the power section, switchgear and to the measuring sensors.

Since commutations circuits are not used in the GTO thyristor converters, and in view of the differing operating conditions as compared to thyristor-controlled converters, the traction control units were modified as follows:

▷ the traction control unit for the GTO converter does not employ any control loops for commutation. For this reason, the tasks performed by the "pulse monitoring" card were considerably simplified;

▷ the pulse technology required for the GTO thyristor resulted in modification of the interface between the traction control unit in the power section.

In the case of thyristor-controlled converters, the energy required for switching must be transmitted together with the control signal. In the case of GTO thyristor converters, the traction control unit supplies only the control signal. The pulse energy for the gate units, which are directly allocated to the phase modules, is supplied by a separate power supply. As a result, the pulse amplifier card subrack is not required in the traction control unit. Fig. 2 shows this design.

Control functional sequence (Fig. 28)

The set-points for traction force, braking force and driving direction are received by the traction control unit by means of a serial computer interface. The traction control unit transmits the current status to the train control unit.

The received set-point values are converted into set-points for the converter and machine controls by means of reference value processing. Limiting and release functions are also active.

Control of the 4-quadrant power controller (4QS)

A specific terminal voltage is required at the input of the 4-quadrant power controller for each operating condition. Based on the assumption that the line currents are of sinewave form, this applies to the secondary voltage of the transformer, taking resistive and inductive voltage drops into consideration, as shown in Fig. 29.

The amplitude and phase of the line-side current with respect to the line-side voltage must be known for this purpose.

The line current is calculated from the constant DC-link voltage and the desired power factor.

The set-point thus calculated for power controller voltage is converted, by means of a line synchronous delta voltage, into a pulse width modulated gate trigger signal pattern for the phase control cards of the 4-quadrant power controller.

The 4-quadrant power controller allows deliberate influencing of specific harmonics in the line current. Use is made of this facility for 100 Hz control. Since the 100 Hz components of the line current interfere with track section release circuits, it is necessary to suppress this frequency.

Practically all control and monitoring functions are implemented by means of software. Hardware circuits are employed in a limited number of cases, in order to reduce the load on the computer.

Machine control

Control of the induction motors employs the principle of field orientation (transvector control). This tried-and-tested technology provides separate control channels for torque and flux of the machine, and is thus comparable with a dc machine, as regards the control capability.

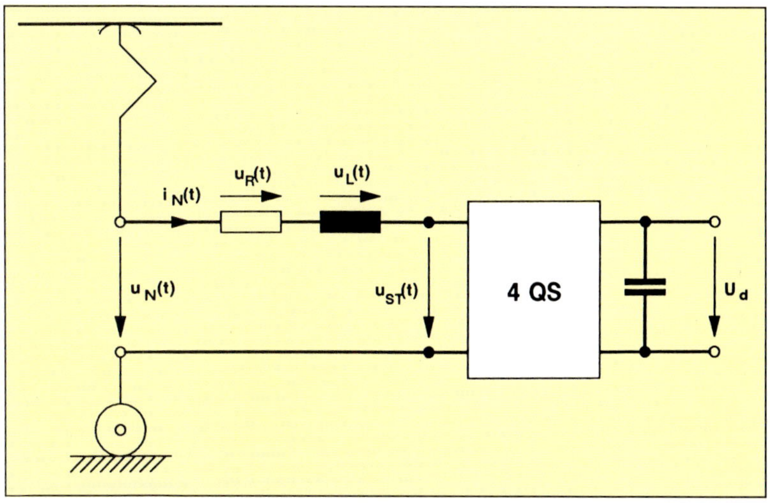

Fig. 29: Equivalent circuit diagram for 4QS

It is of particular significance that this control technique functions independently of critical machine parameters, and does not necessitate precision speed measurement.

Inverter control

The gate signal for the three phases of the pulse width modulated inverter are generated by the inverter trigger unit, on the basis of the frequency and voltage values specified by the machine controls.

This task is performed by a sub-computer. Data transmission between the central computer and the sub-computer is effected by means of a dual port RAM, which can be written and read by both computers.

The trigger unit generates a pulse pattern by means of a reference voltage. A number of different pulse patterns are employed; at low frequencies the asynchronous sinewave sub-harmonic method is employed. Above a defined frequency, synchronous systems with a constant ratio between clock frequency and fundamental frequency, must be employed.

At maximum control signal, a transition is made to fundamental frequency clocking.

The pulses are suitably modulated in order to avoid

sured values from the electrical equipment and the right-hand display values from the braking system of the ICE train.

Fault information, which was previously displayed by means of fault annunciation windows, has been replaced by a diagnostic display for the train driver.

Counter-measures to achieve the function availability are also displayed on request at the push of a button.

All diagnostic data stored in the DAVID diagnostic system can be output in the form of an overview by pushing a seperate button. The diagnostic events are listed in chronological sequence, starting with the earliest event and listing the location, date, time and fault designation.

Further functions

Automatic program preparation

In addition to diagnostics, the DAVID diagnostic system also performs automatic programme preparation. This encompasses both automatic arming and disarming of running/braking circuits, etc.

The times for driver take-over are set in the display before the power unit is vacated. Automatic programme preparation is of particular significance in conjunction with the intermediate coaches.

Bogie monitoring

An acceleration sensor is installed in each of the two power unit bogies. The output signals are amplitude limit value monitored within a frequency range of 2 to 8 Hz for a defined number of cycles. Deviation from the limit value leads to generation of diagnostic data.

Bearing temperature monitoring (LTU)

The following bearings, which are fitted with Pt 100 resistance temperature detectors, are monitored by the DAVID system:

▷ traction motor bearing;
▷ hollow shaft bearing;
▷ gearing intermediate wheel bearing.

The measured values are monitored for violation of temperature limit values. On deviation from a limit value, the diagnostic data are displayed to the locomotive driver who can request display of corresponding countermeasures.

Calculation of braking weight

The braking weight is calculated and displayed on the right-hand display, by pushing a button. For this purpose checkback signals are evaluated from the braking equipment, e.g. brake fault and brake switched off.

Equipment layout (Fig. 34)

Machinery compartment/roof

The various items of equipment in the machinery compartment are arranged symmetrically around the center aisle. The converters are located in the central area. The power cables are routed from the transfromer installed below the floor, to the isolating switches in frames 2.1 and 2.2; then to the inverter units allocated to each bogie, and from there to the

Fig. 34: Equipment layout in ICExpress

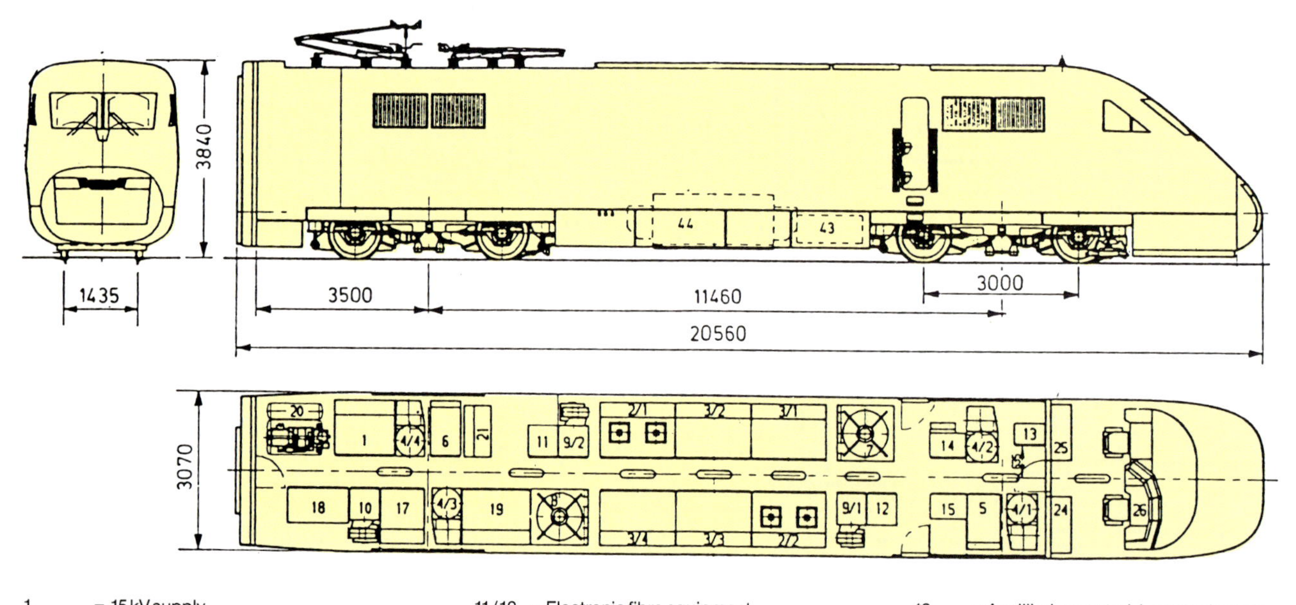

1	= 15 kV supply	11/12	= Electronic fibre equipment, train control	19	= Auxilliaries control, battery charger, fuses
2/..	= Capacitors for series resonant circuit	13	= Optical fibre equipment, train control	20	= Compressed air rack
3/..	= Main converter unit	14	= Protective relays, deadman control climate control, central distance-speed control, diagnostics with automatic train preparation	21	= Brake panel
4/..	= Traction motor fans, motor chokes	15	= Continuous inductive ATC, inductive train protection	24	= Annunciator panel, refrigerator
5, 6	= Bridging contactors, balancing chokes, electronic unit	17	= 110 V control switchboard	25	= m.c.b. panel, graphic timetables, fire extinguishers
7, 8	= Oil cooler with fan	18	= Train busbar, public address system	26	= Driver's console
9/...10	= Converter units for auxiliaries			43	= Battery
				44	= Main transformer

traction motors. This lay-out corresponds to the energy flow and thus results in short cable lenghts.

In accordance with current practice, the power cables, the control cables and the electronic equipment cables have been separately routed, in order to avoid electromagnetic interference.

The four traction motor ventilators draw in air at the side walls of the power unit. They are designed for a delivery rate of 2.1 m³/s. 1.1 m³/s is required for forced ventilation of the traction motors. 0.45 m³/s is required for removal of heat losses of the motor series reactor and 0.55 m³/s for cooling of the machine compartment.

Two cooling towers are provided for removal of the heat losses of the main transformer and of the traction converters. Each tower incorporates two separate oil/air heat exchangers for the design variant employing thyristor-controlled converters. On the basis of extensive tests carried out on the IC-Experimental, it was found that the optimal location for air intake is on the roof.

In the case of power units equipped with GTO thyristor converters, no oil cooling circuit is required. Cooling is effected by taking a partial air stream from the oil cooling tower and routing it through the GTO thyristor phase gate unit cards and the discharge resistors, situated below the floor.

The technical data for the cooling tower in conjunction with thyristor-controlled converters are listed in the margin table:

The main transformer, which weighs 10 t, is installed in the under-floor areas between the two bogies. The transformer and the filter reactor are installed in a common vessel.

The roof carries only the pantograph and the high-voltage penetration. On some vehicles a second pantograph for the Swiss railway system is mounted in addition to the DB pantograph. All other high-voltage components are mounted together with the vacuum circuit-breaker in frame 1 in the machinery compartment. As a result, aerodynamic and aeroacoustic interference, which would be inappropriate in a high-speed train, was prevented.

Driving cab/driver's console

The entrance to the machinery compartment was moved to the vehicle center, in contrast to the IC Experimental. As a result, it was possible to relocate the driving cab rear wall adjacent to the first traction motor ventilator. A spacious driving cab is thus obtained. An extremely userfriendly workplace has been provided for the power unit driver. The lay-out and equipment arrangement in the driver's console was investigated on a full-sized model, in cooperation with driving personnel. The tried-and-tested separation between operating and monitoring levels was retained. The two information displays, for electrical and pneumatic plant information, are unique.

Driving cab air conditioner

The compact air-conditioning plant is installed in the driver's console, as it was in the class 120 locomotive. The air-conditioner is employed for heating and cooling. The cooling output is controlled by reheating cooled supply air. The heating output is controlled by means of an electronic regulator using a 200 v, 16⅔ Hz heater. The compressor and evaporator ventilator are operated at 3 × 440 v, 60 Hz, while the fresh air ventilator operates at 220 v, 50 Hz.

In normal operation, the air-conditioning plant processes circulating air from the driving cab. A fresh air component of approx 15% is fed to the air-conditioner by means of pressure protection devices. Exhaust air also passes through pressure protection devices. Under fault conditions, the driver can effect a changeover to ventilation using 100% fresh air, thereby by-passing the pressure protection devices. The supply air is routed through a duct in the driver's console to the air supply openings at the lower edge of the window.

The cooling air for the evaporator flows from the front end of the power unit to the air-conditioner and is then deflected downwards. The area through which the evaporator cooling air flows within the air-conditioner is sealed to supply air and thus from the driving cab.

Weight of complete dried plant: 550 kg		
Technical data per cooling plant	Converter	Transformer
▷ Heat losses to be removed	155 kW	250 kW
▷ Circulation oil quantity	45 000 litres/h	45 000 litres/h
▷ Oil-side pressure drop	approx 0.4 bar	approx. 0.6 bar
▷ Oil inlet	69°C	97°C
▷ Oil outlet	62°C	86°C
▷ Oil contents (cooler block)	approx 50 litres	approx 42 litres
▷ Max. outdoor air temperature	40°C	25°C
Air-side pressure drop:	stat. 715 Pa	
Air-volume flow:	10.7 m³/s	
Volume (compensation vessel):	approx 210 litres	

Outlook

The state-of-the-art technology installed in the ICE power units represents the most modern implementation of drive technology for railway systems. The first power units of the class are at present undergoing commissioning and were track-tested for the first time in January 1990.

All measures necessary to integrate the new high-speed trains smoothly, when the new timetable is introduced in Spring 1991, have been initiated. It is assumed that all the high expectations of the railway system operators will be adequately fulfilled when the trainsets are actually in operation. The results of operation experience will be presented in the form of a forthcoming report.

□

Sven Bartels and Werner Herbert

High-speed pantograph for the InterCityExpress

Passenger transport of the future requires considerably higher speed if it wants to remain attractive in the face of competition from the airliner and the automobile. The German Federal Railway (DB) has therefore developed a wheel-on-rail system to meet the requirement of "half as fast as an airliner, twice as fast as a private car". Therefore, it was necessary to validate the individual components of the wheel-on-rail system for a design speed of 350 kmph. The pantograph is one of the key components.

The former wheel-on-rail speed world record of 380 kmph was established by the French TGV in 1981 and was limited at that time by the pantograph/overhead wire system. With the Dornier DSA 350 pantograph, a new speed record of 406,9 kmph was established on 1st May 1988, on the new DB track between Wuerzburg and Fulda. This time it was not the pantograph which limited a further speed increase, but the power of the power units.

Fig. 1: General view

The ICE/V (InterCity Experimental) has a power unit each at the front and at the end of the train, plus a variable number of centre trailers. With nine centre trailers, the train has a length of approximately 280 m and provides 432 seats.

The two power units generate an average total driving power of 7000 kW. The ICE/V test train is composed of two power units and three centre trailers and is used essentially for component testing. The success of the development could thus be demonstrated. 60 ICE trains have been ordered so far. Regular service on the newly built route is to start in 1991.

Task definition

The task was defined essentially by the performance specifications of the German Federal Railway.

These are linked to the following requirements:

▷ high-speed running at up to 350 kmph on new lines (RE 250 catenary equipment),

▷ standard speed 220 kmph on upgraded lines (RE 200, RE 160).

Apart from the usual requirements like low wear, reliable operation, low operating costs, maintenance with existing facilities and expertise, the main criterion for the design of a new current collector is the constant contact of the contact strips with the contact wire. Reliable current transmission with this design, therefore, requires as an essential parameter a contact pressure as constant as possible and a standard variation as small as possible.

These requirements must be fulfilled over the complete speed range, for different operating heights, and for different air flow conditions at the pantograph. An active solution with force sensors and regulators was excluded; the problem had to be solved by purely passive methods.

When defining the task, DB planned single-pantograph operation and a 15 kv cable, running through the whole train, to distribute the electrical power between the two power units.

▷ Based on this assumption, a standard deviation of less than 24 N was specified for a given maximum contact pressure of 120 N between the pantograph and contact wire, and;

▷ the folded pantograph was to be as low as possible, which led to certain collateral design conditions.

As the overhead equipment was supplied by DB and was already in use, modifications could be introduced only at the pantograph level, although this was essentially influenced by the overhead wire design. It was therefore, necessary to optimize the dynamic and aerodynamic system.

Dynamic optimization

In the first place, the dynamic behaviour of the overhead wire had to be studied in order to establish a basis for the optimization of the pantograph-overhead wire system with its constant mechanical interactions. The optimization calculations were based on earlier calculations. For this purpose, six spans of equipment of the RE 250 catenary equipment were studied and the interaction between contact wire and pantograph was calculated. After the dynamic behaviour of the overhead wire had thus been determined, the pantograph was theoretically described with a three-mass system. The catenary equipment and the pantograph were integrated through the contact force into a complete dynamic system.

The design requirements resulting from the calculation can be summarised as follows:

▷ lowest possible bow mass,

▷ least possible bow damping,

▷ bow rigidity less than that of the scissors.

The calculation results also showed that a maximum speed of 350 kmph can be reached with this pantograph, in combination with the RE 250 catenary equipment, within the given limit values.

Aerodynamic optimization

For the specified constant contact pressure of the pantograph against the contact wire, the aerodynamic behaviour of the pantograph is of decisive importance. The following requirements have therefore been set up for the system:

▷ low aerodynamic drag in the folded position;

▷ low aerodynamic drag in the operating position;

▷ as far as possible, constant aerodynamic behaviour in different air flow conditions (sideways running, forward running);

▷ low turbulence;

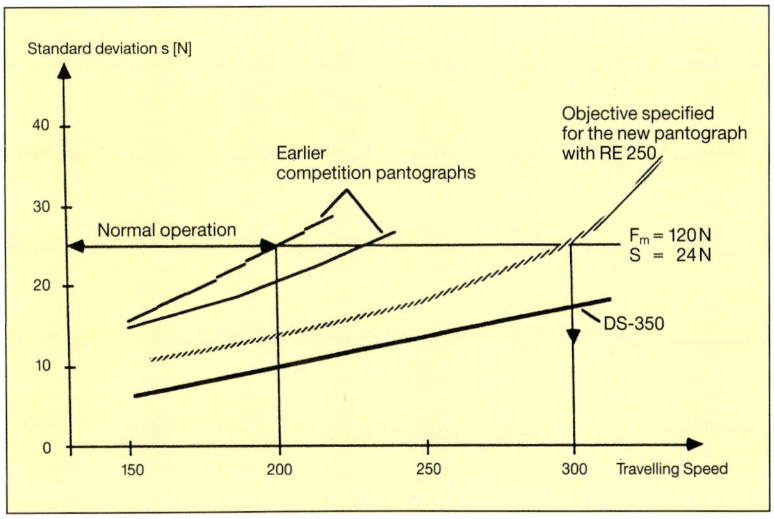

Fig. 2: Standard deviation of earlier pantographs, development objectives, and the result objectives, and the result obtained with the DSA-350

▷ contact pressure to overhead wire less than 120 N;

▷ resonant frequencies of pantograph and overhead wire must not be influenced by induced eddy frequencies.

Optimization results

The development work was closed with a design which yielded a hitherto unknown positive result, even with two pantographs in operation. As no pantograph with integrated insulators was required, the DSA-350 S was developed. The essential optimization results for bow and scissors were fully introduced. The whole design was optimized for production purposes which reduced the production cost by a quarter.

Qualification tests have now been carried out with the DSA-350 S on the ICE and the E 120 electric locomotive on upgraded lines and new lines. It was demonstrated that this pantograph development is

Fig. 3: Optimized DSA 350. This pantograph contributed to the world record in 1988. The insulators are integrated into the structure to obtain a minimum height in the folded position and thus a low aerodynamic drag.

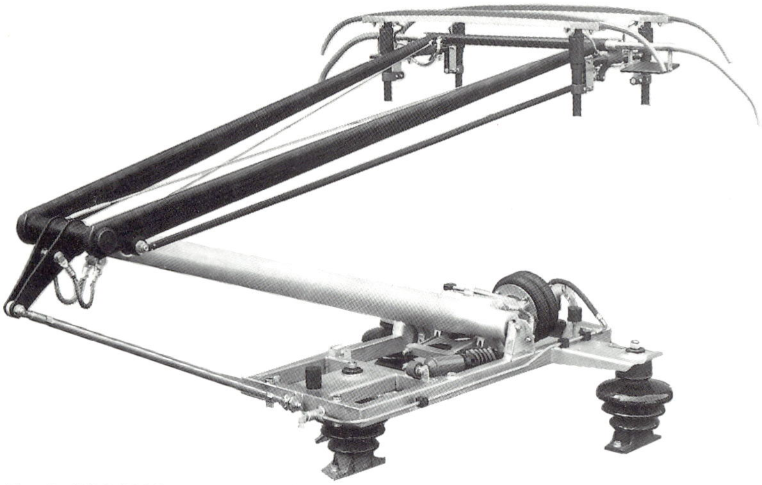

Fig. 4: DSA 350 S pantograph. As a result of the Dornier development, the standard deviation of the contact force was clearly below the specification of the German Federal Railway. Owing to this low standard deviation, it was possible to operate with two pantographs up to a speed of 300 kmph on open line and in tunnel, as well as in different air flow conditions. Even the second pantograph operates nearly spark free. This makes the expensive 15 kv connecting cable between the power units redundant.

an important step forward towards future rail-on-wheel fast traffic, of which the higher speed potential is not yet fully utilized.

Multi-system DSA 350 S

In addition, the requirement of "multisystem suitability" (one locomotive type for the whole European railway network) was taken into account.

A modular concept was chosen for suitability for most of the various voltages and geometries, with a minimum of different components. The mechanical parts can be changed quickly, i.e. the assembly of frame, raising mechanism, lower arm, scissors and bow requires only 10 minutes.

There will be no need of adjustment for weight; low maintenance, no need for painting, insulated ball bearings for highest current in the structure, and few types of screw and nut are further advantages.

A failure detection system (pneumatically operated) will lower the pantograph quickly in case of damage or inadmissable wear of the contact strips. The new type of contact strip will need no maintenance of the horns, because its one-piece aluminium support, including the horn, is carbon-coated.

The adaptation to a.c. and d.c. overhead contact systems will only require a change of contact strips and bondings and the bow springs under some catenaries.

Christian R. Günther and Wolfgang von Waldstätten

Centre trailers

Concept description

General layout

The ICE is a motor train unit with one power unit each at the head and the end of the train and separate coaches between the power units.

In its normal configuration for the German Federal Railway, the ICE has 12 centre trailers, including:

▷ 4 first-class coaches;

▷ 6 second-class coaches;

▷ 1 coach with special equipment;

and one service/restaurant car.

Depending on the passenger volume, varying the number of coaches per train between 7 and 14 units is envisaged.

Fig. 1 shows the centre trailers of the ICExpress The coaches are characterized by the continuous window strip and the closed underfloor pan.

The separate coach principle was chosen for the ICE centre trailer concept, because:

▷ the width of the coach can be maximized for a given coach length;

▷ the train can be adapted simply and rapidly to the

Fig. 1: ICExpress trailer cars

seasonally varying and line-dependent passenger volume, meaning more economic operation;

▷ in case one coach is defective, it can be rapidly replaced by another one, instead of a whole train unit having to be taken out of service.

This leads to greater cost effectiveness, and fewer spare vehicles need to be kept in reserve.

Fig. 2 summarizes some main characteristics of the centre trailers.

Main Characteristics

- Autarc Intermediate Coaches
- Cabin Pressure Control:
  - Air Tightness
  - Special Components
- Aerodynamic Shaping:
  - Closed Underfloor Room
  - Externally Flush Windows
- Extruded Aluminium Body Structure
- Changeable Interior
- Comfort and Communication

Fig. 2: Main characteristics of the centre trailers

Main Data	
Overall Length	26.4 m
Max. Width	3.02 m
Max. Height Passenger Car Restaurant Car	3.840 m 4.295 m
Wheel Base	2.50 m
Max. Axle Load (Restaurant Car)	15.2 to
Number of Seats 1st Class Car 2nd Class Car	 48 66
Max. Cabin-Noise Level	65 dB(A)
Bogie Type	MD 530

Fig. 3: Main data of the ICE centre trailers

Fig. 3 gives the main data of the centre trailers.

With a coach length of 26.4 m for the ICE centre trailers, an optimum compromise has been reached between as long a unit as possible for reasons of economy, and the requirement for as wide a coach as possible (3.02 m) for reasons of comfort, and a car body eigenfrequency of approx. 10 Hz, which decreases as the coach length increases.

The number of seats depends on the level of comfort and space that passengers should be offered.

The German Federal Railway's stringent comfort requirements determine the number of seats in the current set-up.

The number of seats can naturally be increased significantly by decreasing the space between the seats or by adopting the open saloon seating arrangement.

The maximum axle load, which applies to the restaurant car, is still far below the permissible axle load of 20 t for national German and 17 t for international European traffic.

Concept

With the Intercity Experimental a development approach has been initiated for high-speed traffic in West Germany for the early Nineties and, on some lines, also for border-crossing traffic. This approach was applied initially in concrete terms to production vehicles developed for the ICExpress as described here. Advanced development in the form of the border-crossing ICE-M multiple unit has already been launched.

The concept for the centre trailer of the multiple unit for the ICExpress complies in the first instance with the requirements laid down for the individual vehicles making up the whole train as a self-contained unit. The concept elaborated by the German Federal Railway for high-speed, long-distance transport is based on principles which have been consistently developed since the adoption of the two-class Intercity system in 1979.

The layout and the interior design of the individual ICE coaches are shown in Figs. 4 to 11.

The interior design, furnishings and coach combination are described below.

Seating with two levels of comfort, a catering service in the train, separation of the two levels of comfort by the service car and various seating arrangements in the form of open saloons and ordinary compartments are the basics that determine the number of coach types and their sequence in the train unit. Further services that will in future constitute an important part of long-distance passenger traffic include, for instance, a conference room and one telephone booth each in the two coach groups of the different levels of comfort. The disabled toilet and two parking spaces for wheelchairs, which are already familiar from Intercity trains, are also features of the ICExpress.

The differing wishes of passengers for seats in normal compartments, in open saloons and in rows with tables in between were complied with by offering approximately the same number of all three layouts in each coach.

The differences between the two levels of comfort are evident from the distance between the seats, from the compartment width, and from the seating arrangement in the open saloon and in the areas with tables. The superior level of comfort features a seating arrangement of 2+1, and an arrangement of 2+2 in the Second Class.

The service coach plays a special role in the multiple unit, not least because it separates the two levels of comfort in the train. This separation has been continued right up to and including the service coach – the latter offering catering services in two different price categories. A development approach is continued here which began with the adoption of the on-board restaurant in Intercity trains, i.e. dining car with a novel catering concept. In line with this concept, the new service coach in the ICExpress should resemble the on-board restaurant in Intercity trains by providing, despite its separating function, a cen-

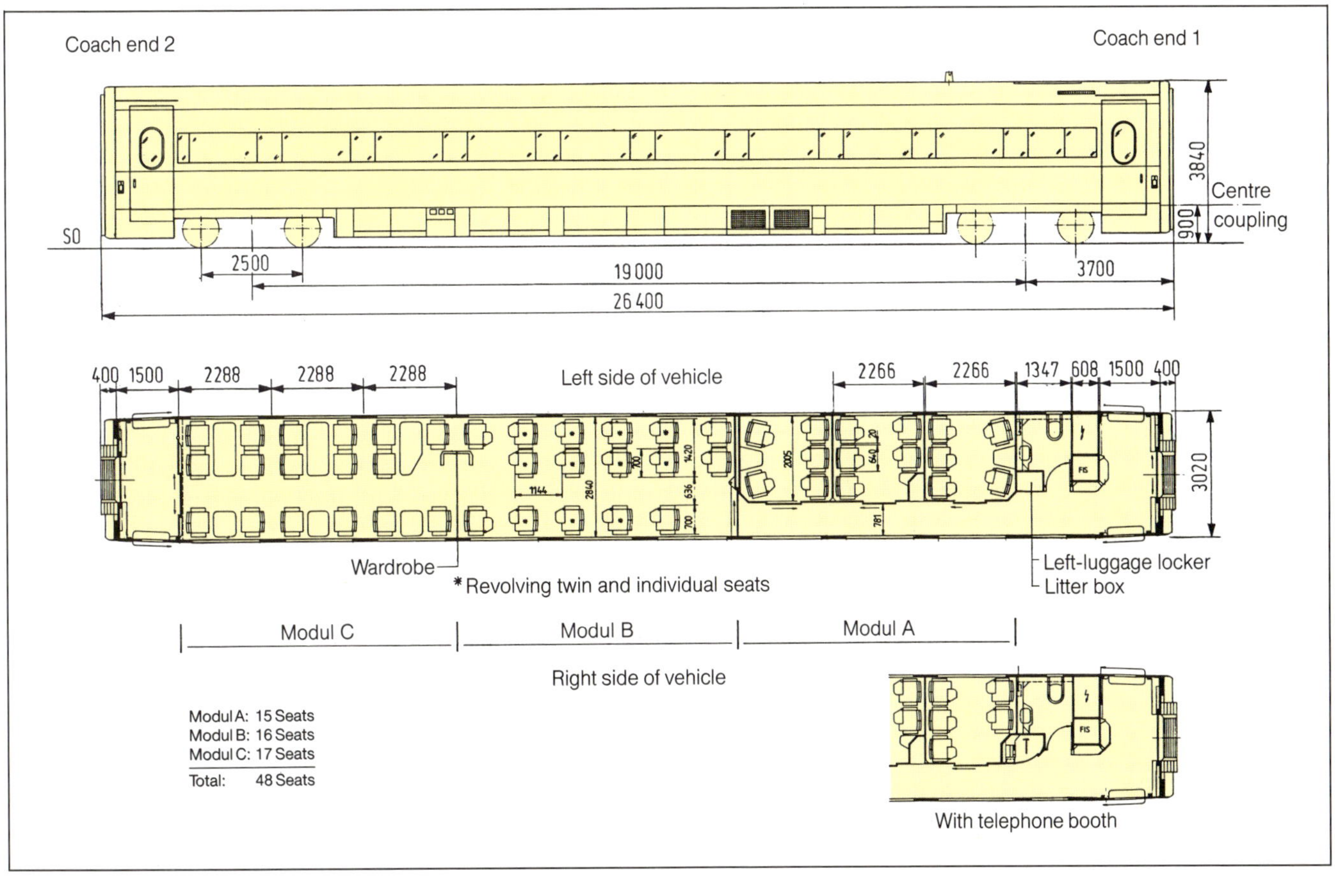

Fig. 4: Layout of the ICE 1st class centre trailers

tral meeting point in the train and thus a place of communication.

Special equipment that need not be duplicated in the multiple unit is concentrated in one coach, and comprises mainly the facilities for handicapped passengers, a conference room, and compartments for train attendants and service personnel. The Passenger Information System Centre is also to be found in the train attendants' compartment. The coach with the special equipment together with the service coach in this way form the nucleus of the ICExpress multiple unit. The coach with the special equipment is Second-Class, the special equipment replacing the compartments.

Fig. 5: Business travellers in a 1st class saloon with of a type Avmz 801 centre car

Fig. 6: Travellers in a 1st class saloon of a type Avmz 801 centre car

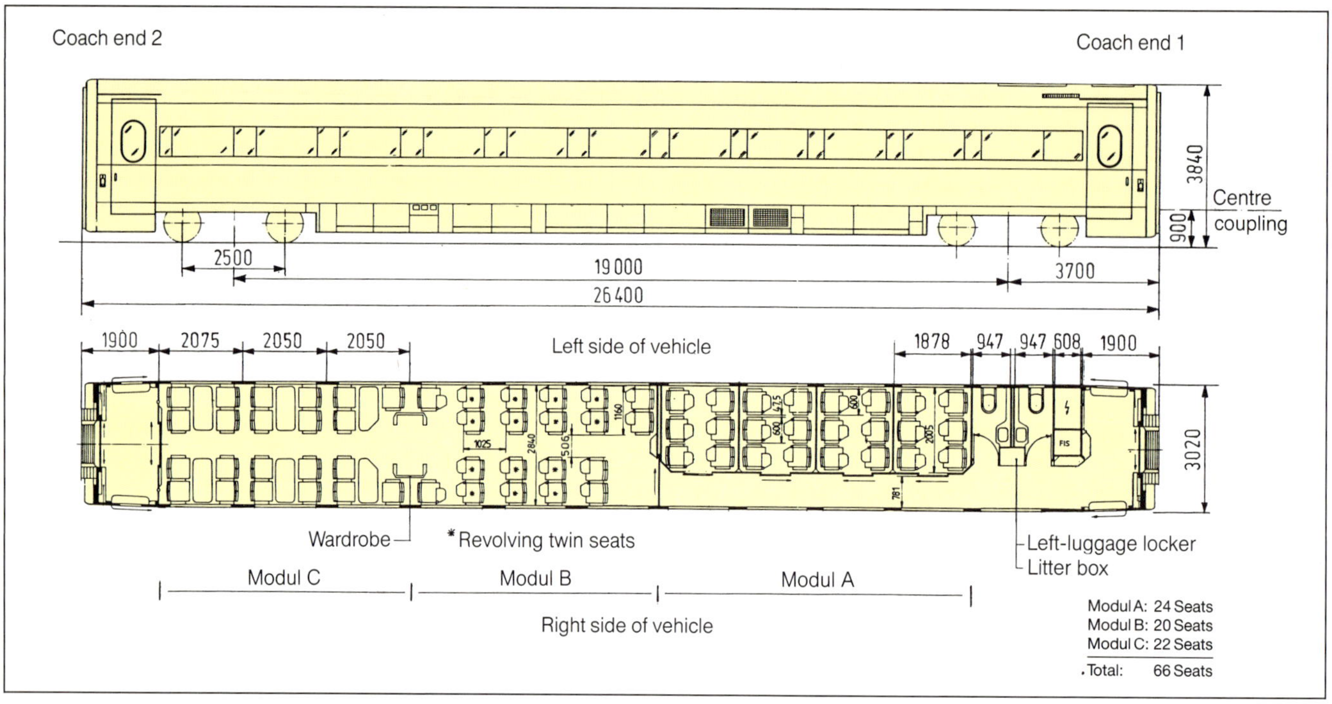

Fig. 7: Layout of the ICE 2nd class centre trailers

All the services for passengers in the ICExpress multiple unit have thus been made available in only four different types of coach.

Vehicle structure and components

The main dimensions of the ICExpress centre trailers were established with a length of 26.4 m over buffer and a maximum width of 3.02 m. These dimensions can be implemented without any problem on the German Federal Railway's high-speed line network (which is at present under construction or at the planning stage) on most connecting and passing lines. Border-crossing traffic into the countries using the same current network as the German Federal Railway is also feasible.

The coach bodies for the ICExpress centre trailers are made according to the tried-and-tested aluminium extruded-profile construction method. Optimization was achieved allowing for the maximum cost-effective profile size, the expenditure for thermal welding methods, and the resulting reworking.

By concentrating the technical equipment at one end of the coach, it was possible to devise almost uniform shell designs, which differed only in the dis-

Fig. 8: Layout of the ICE 2nd class centre trailer with special compartments

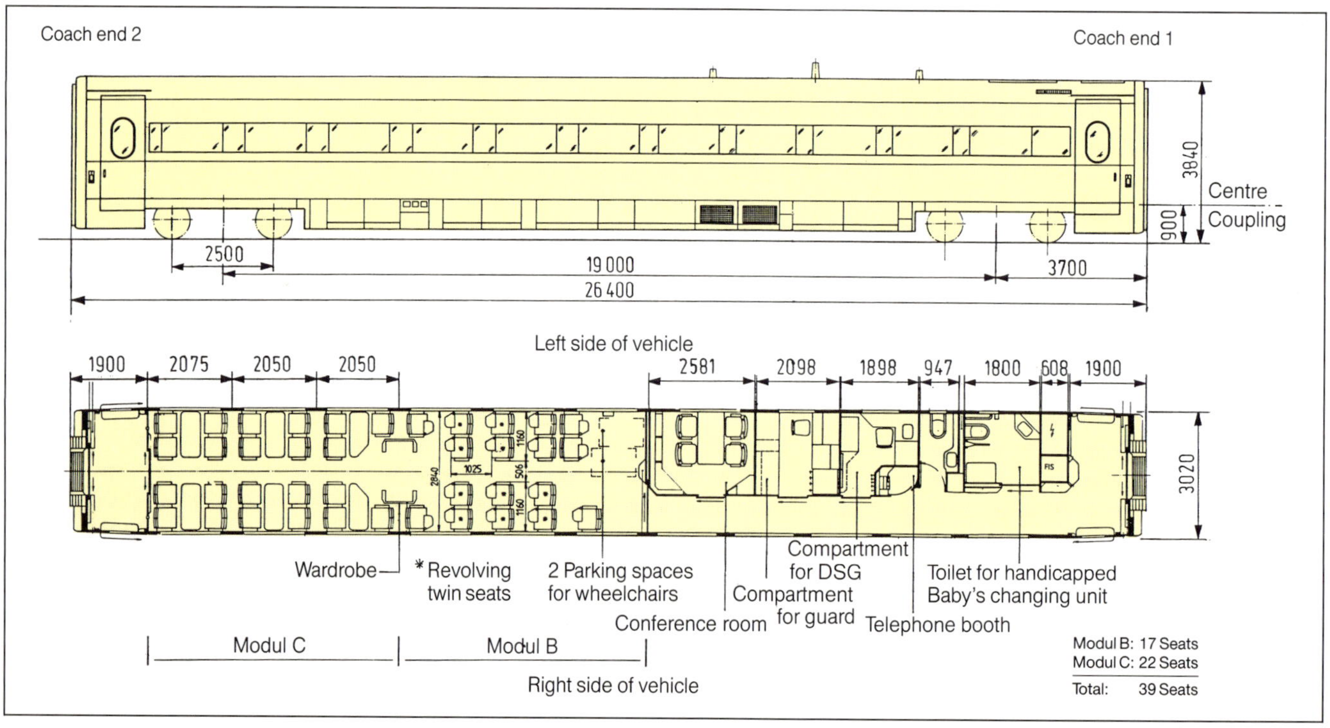

tances between the windows, the size and position of the services windows, and the arrangement of the services for the WC systems.

A special solution had to be found for the service coach. The regulation minimum height for the kitchen section and additional technical essentials made it necessary to make this coach higher than the passenger coaches. The external design of the service coach was devised in line with these conditions to become a special feature of the train unit.

The transition from the lower roof of the passenger coaches to the top of the roof of the service coach, roughly 455 mm higher, was implemented by means of roof slopes designed at angles of about 16 degrees.

Roof windows in the passenger areas serve to highlight the special features of this coach to make them more conspicuous.

The coach body shells for the ICExpress centre trailers have been pressure-sealed. Besides the careful construction of all necessary openings and the sealing of all unwanted ones in the car body structure, this includes designing the body shells with allowance for the anticipated aerodynamic loads.

Fig. 9 (top): Travellers in a 2nd class saloon of a type Bvmz 802 centre car

Fig. 10 (left): Travellers boarding a ICE-train; railway policemen assist a handicap; (centre): 2nd class centre coach BSmz 803 with handicap toilet; (right): Conference compartment with telephone, type-writer, copier and telefax

All the centre trailers for the ICExpress feature a closed underfloor pan accommodating the brake aggregates and major parts of the air conditioning plant and electric power supply as well as the non-drinking water and sewage systems. These systems are accessible for maintenance through hinged flaps. The movable side skirts are made of aluminium extruded profiles, and the plane floor sections are designed as lightweight sandwich panels coated with Nirosta sheeting as a protection against damage from rockfall.

The latest materials developments have been taken into account for the heat and sound insulation systems. All the insulation materials used in the passenger sections, side walls and ceiling are non-combustible and comply with more stringent safety requirements, particularly in respect of the recently-built long tunnel sections.

The concept of the continuous window strip with viewing and blank windows flush with the skin, which has, in the meantime, become a feature of ICE vehicles, was taken over for the centre trailers of the IC Express. The sound-insulation values were improved considerably by modifying the window structure. A blind that can be individually adjusted manually is fitted in front of all the windows for protection against the sun.

The ICExpress passenger coaches feature two access doors each at both ends of the car. The floor level, 1210 mm above top of rail, is reached from the platform via one folding step and one fixed footstep. The steps have been optimized for the most frequent national and international platform heights of 760, 550 and 380 mm above top of rail.

The end sections of the coaches with access doors, steps and vestibules have been designed identically for all passenger coaches.

The coach doors are single-leaf horizontal swing sliding doors. They are fitted with a double-acting profile rubber sealing system against aerodynamic shock pressure; a safety device to prevent anyone from getting caught in the door is integrated; it reacts to the force of pressure and, when operated causes the door-closing movement to be reversed. The window in the coach door has been designed as an emergency exit with a removable rubber sealing profile.

The service coach in the ICExpress has no coach doors. There is a loading door on each side roughly in the centre of the car. The loading door design was derived directly from the passenger coach doors.

The inter-coach gangway connections comprise continuous double rubber joints and a bridge of moving links. The double rubber joints enclose the coach gangway with the floor plates and the centre buffer coupling located below. The double rubber joints can be separated by a special coupling frame at the coupling level, likewise the bridge of links.

A rubber shroud round the contour of the coach ends to bridge the gap between the ends of two coaches has an aerodynamically favourable effect. The passageway between the power car and the intermediate coaches has been designed so as to be completely identical with the coach passageways.

Fig. 11: Layout of the ICE restaurant car

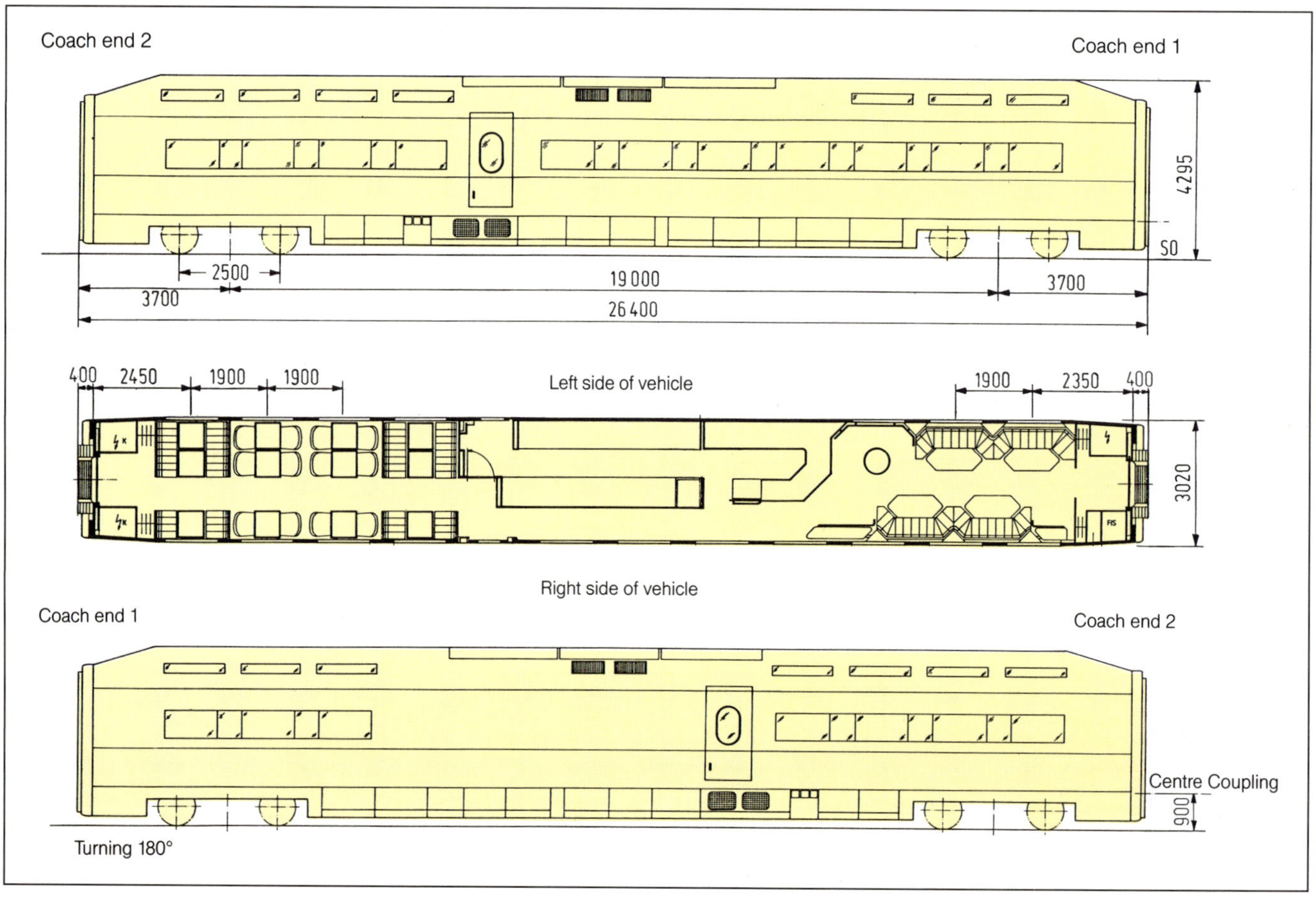

Interior furnishings

The ICExpress centre trailer was designed under the direction of the German Federal Railway's Design Centre at the Munich Central Railway Office. The problem was to devise a homogeneous, consistent design for the three different interior versions of the coach. To tackle this problem, the German Federal Railway made full-scale models of the interiors of each coach representing all the details of the design development, where possible with the original materials.

One order was placed by the Federal German Railway as a separate development for a major element of the interior furnishings. The most important features of this seat development constitute adjustment of the back-rest folding tables in the armrests, foot supports in the 1st Class coaches, and the integration of seat-related fittings for the Passenger Information System.

Special problems, but also special possibilities, became apparent on designing the passenger sectors in the service coach. The roof height, raised for technical reasons, was exploited in the form of a ceiling structure, decisively influencing the impression of space. The light effects of the roof windows play a special additional role.

The two customer sections are based on the catering concept for the ICExpress and have been devised as a restaurant with waiter service, and self-service in the snackbar section. The restaurant is a non-smoking area, while smoking is permitted in the snackbar. The restaurant has 24 seats, 12 being fixed and 12 movable. There is standing room for 10 and 16 seats in the snackbar section. The restaurant is carpeted, and the snackbar and corridor have a two-colour rubber nap-finished floor covering. The service coach is the only one in the ICExpress to have doors at the ends, which remain closed in normal position.

The kitchen and the snackbar counter are located in the centre part of the coach. The kitchen is separated from the restaurant by an electric revolving door and from the snackbar counter by a manually operated sliding door. The central passage in the kitchen is 700 mm wide, and the kitchen equipment is located on both sides of this passage.

The technical equipment in the kitchen has been devised mainly for putting the final touches to the prepared food that is supplied. However, it is also possible to prepare meals conventionally. The meals offered in the restaurant car are beeing adapted increasingly to the wishes and needs of more and more passengers for light, fresh, and healthy nourishment.

Equipment

The air conditioning system for the ICExpress centre trailers was newly devised on the basis of experience with previous vehicles. A large ceiling air inlet with low inlet-air flow rates was designed for the first time in the history of rail vehicle construction. Extracting the circulating air in the upper and the exhaust air in the lower side-wall area provides a good mixture of air inside the coach. Floor heating has been installed in order to prevent a cold-air zone from occurring above the floor. In addition, an electric convection heating system has been fitted under the windows to avoid the cooling effects of the side-wall and window areas.

The air conditioning system requires the provision of apertures in the otherwise pressure-sealed vehicle shell for the intake of fresh air and extraction of the exhaust air.

To protect the passenger area from unacceptably high, and rapid changes in, pressure, air intake and extraction are via appropriately designed and controlled pressure protection ventilators and pressure protection valves.

The air treatment unit of the air conditioning system is located in the roof area above the toilets and installed and removed via a hatch in the roof; the same applies to the pressure protection ventilator for fresh air. The air treatment unit is accessible for maintenance from inside the coach through hatches in the roof. The compressor condenser aggregate is installed in the underfloor pan, and access for maintenance is through the movable side skirts.

The WC system and the non-drinking water and sewage systems features the Semco design that has proved so successful with the German Federal Railway's coaching stock. The water tanks are installed in the underfloor pan and fitted with connections on both sides for fast filling and waste disposal.

The elements of the electric power supply have been designed in the familiar, tried-and-tested way. The major assemblies have been combined in the well-known manner to form a power supply block that is installed in the bogie.

The brake system and its most important assemblies are mounted in the bogie. The brake-operating unit represents a major innovation and is located inside the coach near the electronic equipment cupboard. The controls are no longer accommodated on the external longitudinal member but are now accessible only from the entrance vestibule through a locked door.

Bogies

A bogie has been designed for the intermediate coaches of the ICExpress which is based on the results of the running analyses conducted during line tests and trials on the roller test rig with the ICExperimental bogies. The new bogie is an advanced development of the proven MD 52 bogie, from which the MD 52-350 high-speed bogie was derived. Taking the test results for the MD 52 production bogie into consideration, the bogie wheelbase was defined at 2,500 mm. The new bogie, designated MD 530, was made slightly narrower, in view of the arrangement of small bogie skirts. The bogies feature a double device for yaw damping, and the version for the service coach is equipped with a roll stabilizer. The brake system on the bogie side, consists of four axle-mounted brake discs per wheelset and an electromagnetic rail brake. The bogies, as well as the coaches, are prepared for alternative equipment with a linear eddy-current brake.

Fig. 16 shows a door with locking and sealing elements.

When closed, the doors are flush with the outside contour of the coach body. At a suction of 5,500 Pa, a force of approx. 1 t acts on a single door, and the measurements of the door leaves, the seals and the mountings must be appropriate.

In addition, it should not be forgotten that the doors must fully meet air-pressure requirements, particularly in different weather conditions.

The doors for the ICE production coaches were subjected to comprehensive fatigue tests on a special test rig and in a climatic chamber, with excellent results.

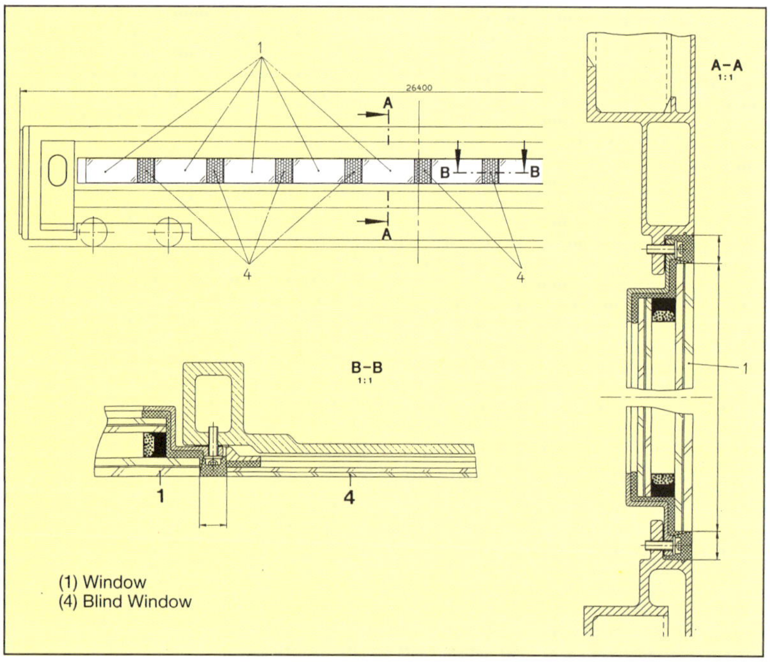

Fig. 17: Cross section of the ICE trailer car windows

Windows

The ICE centre trailers for the Federal German Railway feature a continuous window strip flush with the outside skin and consisting of blank and viewing windows. The blank windows create an attractive design effect.

Fig. 17 shows a cross section through the windows.

The German Federal Railway's stringent requirements for air-pressure and water vapour impermeability, and minimization of aerodynamic drag are met, thanks to a new solution based on bonded window panes flush with the outer skin.

The panes are bonded into aluminium frames. The complete window units are screwed into the coach body, and the joint between adjacent windows and between window and coach body is plugged with an elastic filling material. This provides absolute impermeability to pressure and moisture.

This principle has already been applied to the centre trailers of the ICExperimental with very good results. The windows were subjected to a long series of fatigue tests in various environmental conditions, and passed comfortably.

The static and dynamic design of the windows, must be executed with the utmost care to take into consideration intermittent pressures when in tunnel or when passing trains. For instance, suction forces of approx. 0.8 t act on the largest windows of the ICE when in tunnel. Here the long-wave pressure variations which may occur in tunnels are more unfavourable than short-wave loads occurring when two trains pass each other on the open line.

The function of these windows also necessitates careful preparatory work, tests and quality assurance of bonding, UV protection, simple and rapid as-

Fig. 18: Gangway

sembly and adjustment, as well as simple dismantling if windows have to be removed.

Intercommunication gangways

Conventional intercommunication gangways in use today fall far short of air pressure requirements. Consequently new air pressure-tight gangway connections between coaches had to be developed and built which, in the meantime, have also been installed in other DB vehicles designed for high-speed DB tunnel sections.

The so-called double rubber joints were chosen for the ICE series from various possible solutions. This is a gangway enclosed on all sides and consisting of an outer and an inner rubber joint.

Fig. 18 shows the gangway.

Fig. 19 demonstrates the gangway principle.

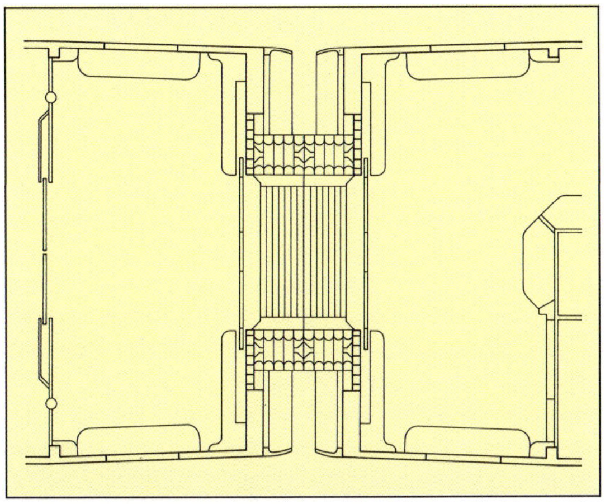

Fig. 19: Pressure protection gangway

One half of the intercommunication gangway is attached to each coach end. After coupling the coaches by means of an automatic coupling, both gangway halves are mechanically locked together.

Baffles are attached to the outside of the coach which, for aerodynamic and design reasons, reduce the gap between the ends of the coaches to the minimum necessary to allow for the negotiation of tight curves. The pressure-tight gangway connection encloses the gangway floor plates and the coupling section.

Toilet system

For reasons of pressure tightness as well as of hygiene and environmental protection, a retention toilet system was chosen for the ICE, which collects the waste in a container to be emptied at appropriate intervals at railway terminals, through automatic suction devices.

Of the possible retention toilet principles, the vacuum toilet was chosen; this has already proved suitable in several vehicles.

Fig. 20 shows the schematic structure of the retention toilet system.

One advantage of this system is its pressure tightness; the waste is evacuated through low pressure. Two staggered valves are switched in sequence.

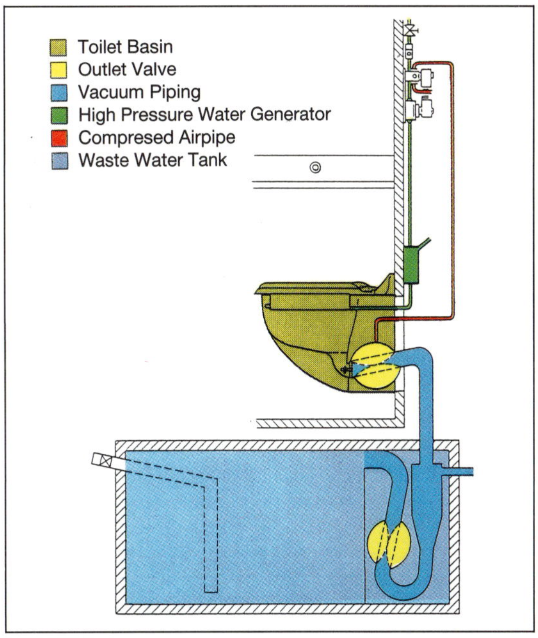

Fig. 20: Retention toilet system of the ICE trailer cars

There is thus no danger of pressure compensation via vent and suction pipes.

A further advantage of this retention system is that it is hygienic – there is no danger of the vehicle being soiled by waste being carried away by wind, or of environmental pollution.

Running behaviour

The main requirements for bogie design are:

▷ minimum curve radius 150 m,
▷ maximum speed 300 kmph,
▷ ride quality limit value 0.5 m/sec² horizontal/vertical
design value 0.2 m/sec² horizontal/vertical
(root mean square values);
▷ wheel profile UIC/ORE S 1002 (worn wheel profile),
▷ safe running behaviour over the entire speed range and on all line elements,
▷ minimum possible load on track, i.e. low lateral forces on track,
▷ high reliability and low maintenance i.e. low wheel wear.

A two-axle bogie with steel springs which has been in use for many years in IC trains run by the German Federal Railway, at speeds of up to 200 kmph, has been further developed for the higher speed range in the light of these requirements, for which air-sprung two-axle bogies are also being developed.

The bogie with steel springs, as well as an air-sprung bogie, were tested in the ICExperimental at speeds of up to 400 kmph with good results.

The advanced development of the bogie with steel springs was chosen for the first series of the ICE trains, in view of long and reliable experience with these bogies in the IC trains operated at speeds of up to 200 kmph by the German Federal Railway.

Fig. 21: MD 530 trailer bogie

Fig. 21 shows this bogie, designated MD 530, for the speed range up to 300 kmph. This is a two-axle bogie with steel springs with a double heating control.

Fig. 22 shows the bogie layout.

At 2,500 mm, the bogie wheelbase is very short so that no space for installation is wasted in the coach body.

The bogie has two brake systems, a disc brake system with, optionally, four brake discs per axle, and an electro-rail brake system. An eddy-current brake may be fitted later.

The number of brake discs depends on the service range and the braking requirements and should be recalculated for Korean needs.

A so-called wear profile is used for the wheels, offering the advantage of a good kilometric performance but calling for extremely well-stabilized bogies.

The quality of the bogies can be demonstrated by means of the favourable curve of the lateral forces.

Fig. 22: MD 530 bogie design

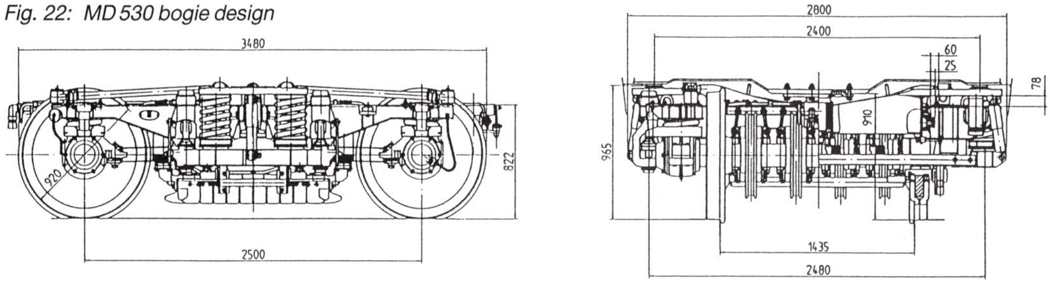

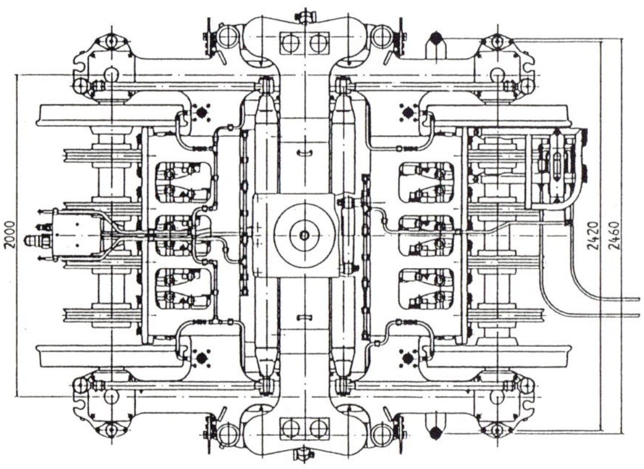

Fig. 23 shows the wheelset guidance forces of test wheelsets, plotted against the speed for the MD 350 bogie, the original bogie installed under the test coach of the ICE Experimental.

The measurements were carried out on the straight section of the high-speed Gütersloh–Neubeckum line at speeds of over 300 kmph.

The two limit curves represent the reliability range with a reliability probability of 0.15 and 98.85 per cent.

Fig. 23: Wheelset guidance forces

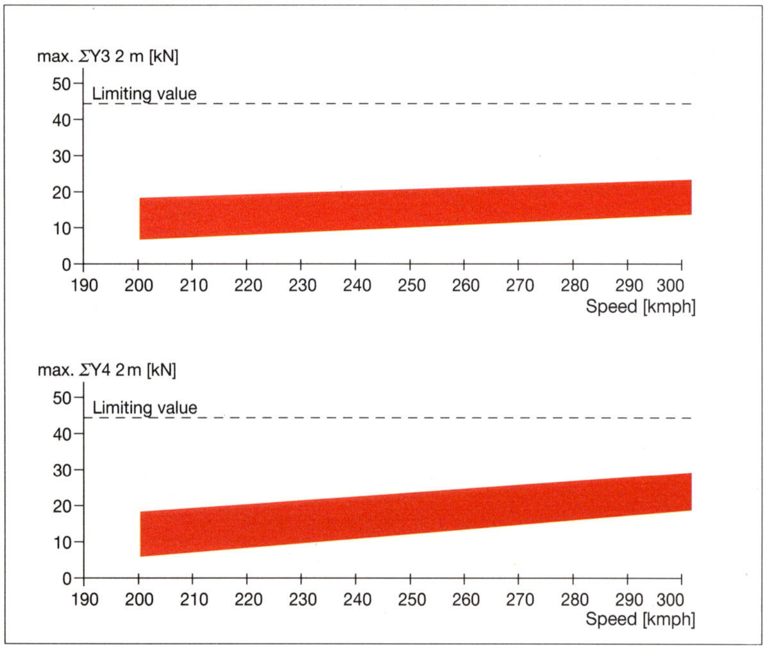

Fig. 24: Bogie with air suspension and fibre composite frame

The test runs showed satisfactory stable running with a low cross-section level. Rides in curves had similarly good results.

The German Federal Railway uses a so-called quality characteristic "Wz" among others as the assessment criterion for the running quality in the coach body.

For reasons of comfort, the limit value is WZ = 3.25.

A Wz of less than 2.5 should be aimed at for the ICE. On the forerunner vehicle, the IC Experimental,

▷ a ride quality of 1.9 to 2.1 laterally;

▷ a ride quality of 2.3 to 2.6 vertically;

were measured at speeds between 300 kmph and 325 kmph.

In addition to the MD-530 bogie with steel springs used in the present ICE series, a number of advanced bogies are available for the high-speed range. Fig. 24 shows, for example, a new air-sprung high-speed bogie.

This bogie has been developed particularly for lightweight construction and low noise, and consequently its frame has been made entirely of fibre composite and also takes over the function of primary suspension. This results in considerable savings in weight and components and in noise reduction.

The bogie shown is currently in service under an Intercity coach of the German Federal Railway and has already covered a test run of 600,000 kilometres without any problems.

Coach body structure

The body of the ICE centre trailer was built of aluminium integral construction with large extruded profiles.

The advantages of this construction method, which led to its selection for the ICE, are:

▷ lower weight compared to steel construction;

▷ smooth surfaces which need be, if at all, only slightly reworked;

▷ stiffer construction compared to lightweight steel construction, so that less coach body pumping movements occur in tunnels;

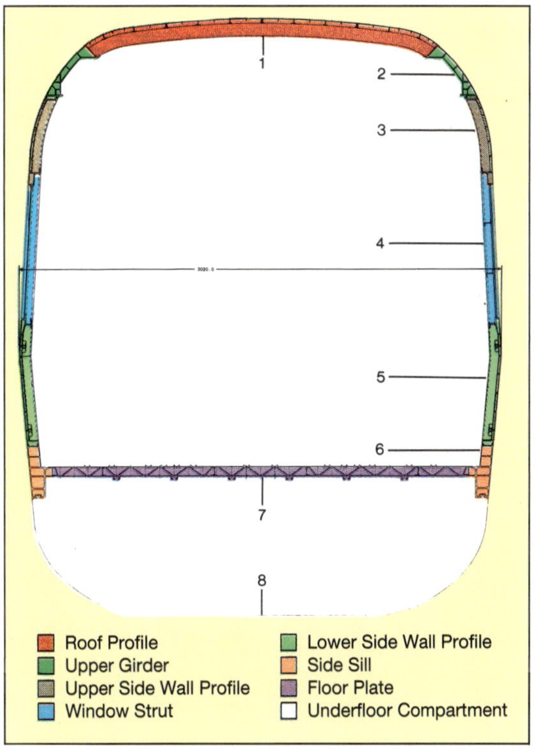

Fig. 25: Coach body structure cross section

▷ easier pressure-sealing of the structure owing to the large size of the components;

▷ reduced susceptibility to corrosion;

▷ simple component attachment owing to the pos-

Fig. 26: ICE-Coach body structure

sibility of the integration of fixing rails, e.g. on the floor and side panels, directly in the extruded parts.

The extruded components for the floor section, side panel and roof are manufactured in one piece over the whole length for the ICE coaches.

The maximum weight of one component of this length is 20 kg/m, and the maximum component width 800 mm.

The minimum wall thickness which can be achieved is 2.5 mm.

Set-up of the shell structure

Fig. 25 shows the cross section of the ICE coach body shell, extruded profiles welded together.

The roof 1 is 3 profiles welded together, and the upper girder 2, the upper side-panel profile 3, the lower side-panel profile 4 and the side sill 5 consist of one profile each. The floor section 6 consists of 5 profiles welded together.

The mounting rails for the equipment are pressed into the lower side of the floor section.

The underfloor pan 7 does not form part of the stressed coach body structure but serves merely to cover the equipment room. It also includes side flaps for equipment installation and maintenance.

Figs. 26 and 27 shows photographs of the shell assembly.

Fig. 27: ICE-Coach body structure

The shell structure for the ICE passenger coach weighs 8.2 tonnes.

The first vertical bending eigenfrequency of the shell is high enough at 15 to 17 Hz depending on the type of coach, to ensure an eigenfrequency of 10 Hz when the shell is completed.

Static and dynamic dimensioning

The pressure load occurring in tunnels and when trains pass each other must also be taken into account for static and dynamic structure design as well as for the customary general design requirements for each train.

Considerable attention must be paid to the careful superposition of weight-related and aerodynamic loads.

Following extensive analyses, we have found a procedure for the ICE that realistically reproduces the superposition cases.

The Finite Element Method was applied to static and dynamic dimensioning of the coach body.

Fig. 28 shows one of the many models of the overall structure.

The vibration response of the structure and of the completed coach body was determined with the aid of modal analysis.

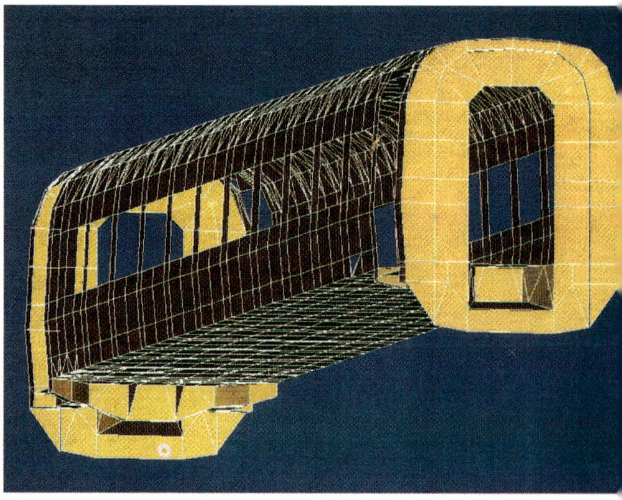

Fig. 28: FEM model overall structure

Expertise

Sound processing expertise is essential to the successful application and utilization of all the advantages offered by aluminium construction. Such expertise is available to the West German industry owing to many years handling this modern material, for instance:

▷ the special welding technique, seam preparation and quality assurance must be mastered;

▷ the special materials properties must be borne in mind for shaping;

▷ postcuring and shrinkage behaviour must be mastered for reproduction.

Train Information System (TIS)

The ICE Passenger Information System (TIS) offers information, entertainment, communication and services to the passengers, while providing support for the crew.

The following installations are linked to form a complete system:

▷ controlled displays;
▷ Btx-type information terminals (Btx is the German Teletext system);
▷ public-address system;
▷ music programmes via headphones;
▷ video installations;
▷ telephones;
▷ driver/crew intercommunication;
▷ paging system (bleepers);
▷ conversion of the European Radio Call;
▷ passenger service-call systems.

TIS is a modular system with a hierarchical structure. It consists of:

▷ a control centre;
▷ subcentres;
▷ terminals.

TIS exploits state-of-the-art CMOS microprocessor technology for the construction of electronic assemblies that are suitable for use in vehicles. Such advanced technologies include, for example, colour LCD and EL (electro-luminescence) for displays.

The use of TIS in the German Federal Railway's Intercity Express train (ICE) illustrates the system's wide range of possibilities and functions. Its modular structure allows for a configuration tailored to the specific requirements of various transport systems.

TIS for the ICE is based on a central control system to which subcentres, one per coach, are linked. These coach control systems, in turn, communicate with the TIS terminal devices. This basic structure for signal transmission uses separate paths for data and for wideband RF transmission.

Figs. 29 and 30 give an overview of the TIS installation in the coach.

Information

The public-address system in the passenger area transmits not only automatic announcements in syn-

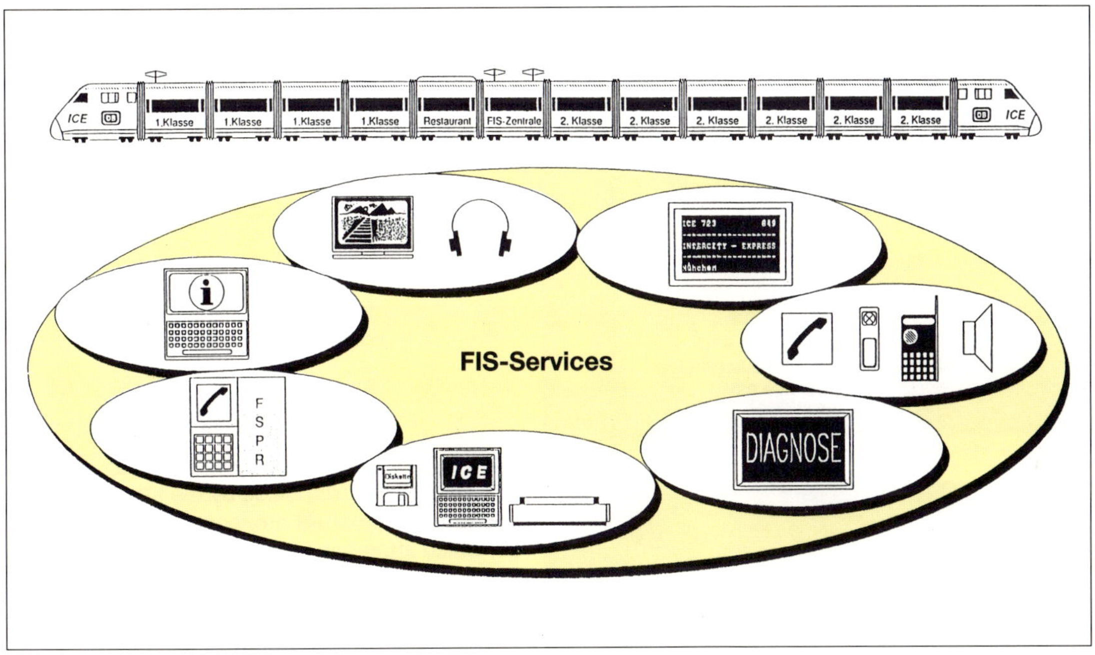

Fig. 29: Passenger information system, services

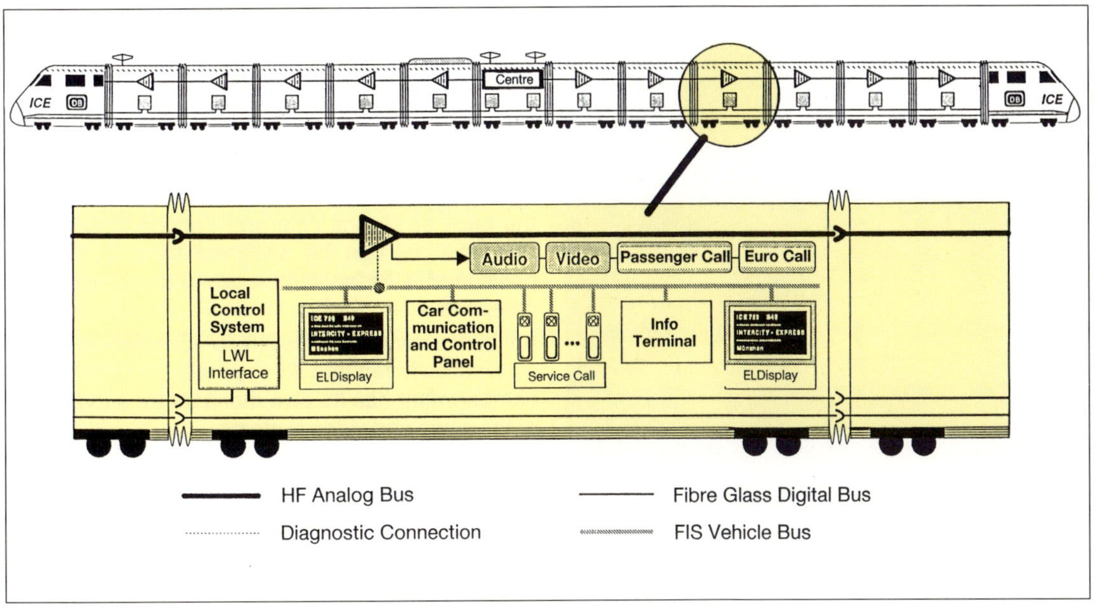

Fig. 30: Passenger information system, coach installation

For winter, floor-section heating has been installed in addition which keeps passengers' feet warm, in contrast with other air-conditioning concepts.

To ensure that window panes are free of ice in extreme wintry conditions, a secondary air-heating system will be installed behind the side panels below the windows, to generate warm air in front of the windows.

Noise level

A minimum noise level is also an essential criterion for a high level of comfort.

Since noise propagation increases approximately at the fourth power of the train speed, special efforts must be made to dampen noise in high-speed vehicles.

One of the most important measures is to reduce noise at its source.

Such measures are:

▷ the use of noise dampers on all wheels.

▷ good aerodynamic coach design with the windows and doors as flush as possible with the outside skin, baffles at the intercommunication gangways, and a fully faired underfloor to keep aerodynamic noise to a minimum.

Measures to dampen coach-body noise consist of:

▷ good noise insulation of the coach body (the most advanced material developments were used here),

▷ good sound insulation of the coach doors and intercommunication gangways,

▷ a window structure ensuring the same noise damping as with the sound-proofed coach body;

▷ and sound damping measures in the ducts of the air conditioning system.

The German Federal Railway's requirements envisage that the noise level in the passenger compartments at 250 kmph must not exceed;

▷ 65 dB (A) in the First Class,

▷ 68 dB (A) in the Second Class;

when in-tunnel the noise level may be 5 dB (A) higher.

At the coach doors and in the vestibules, in the intercommunication gangways and WC's, the noise level must not exceed 75 dB (A) at 250 kmph and in-tunnel.

In view of these measures, we shall generally achieve lower noise levels in the interior of the ICE than those mentioned above.

The ICE will generate less outside noise at 300 kmph than a conventional train at 200 kmph.

Power supply

The concept of the electric power supply has been developed and grouped into modular elements, in view of the previously discussed topics.

The power supply concept for the ICE centre trailers is shown in Fig. 36. (Courtesy KEK-Consortium).

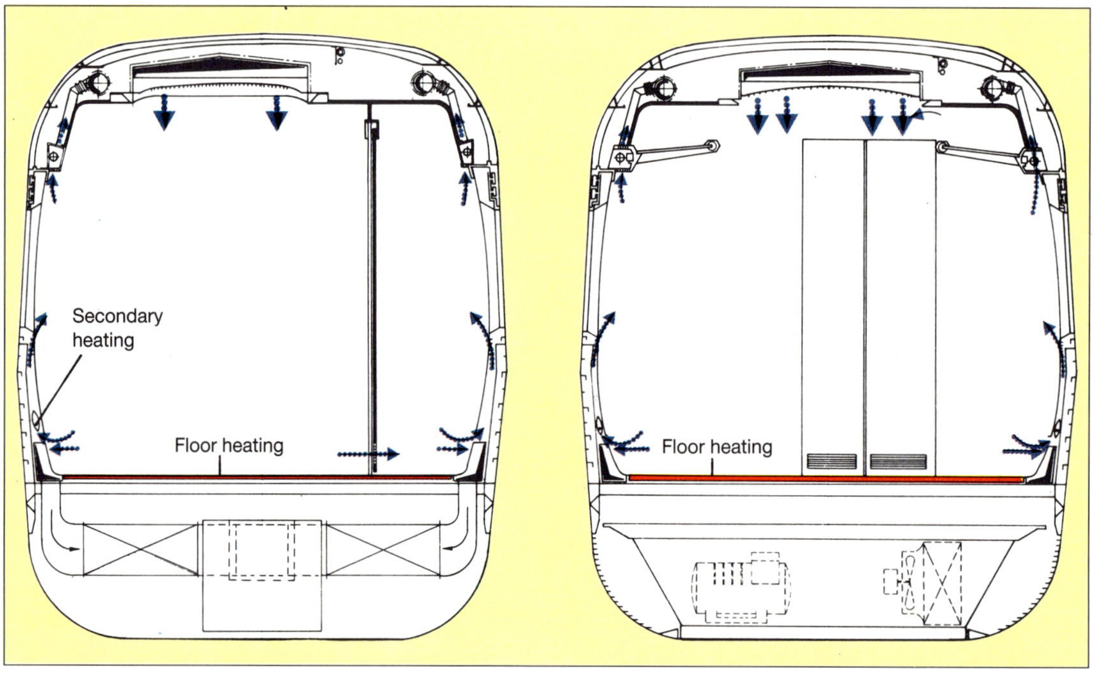

Fig. 35: Air conditioning system

The power supply for the centre trailers is ensured via two train bus bars with 1,000 v, 16⅔ Hz. Each power unit supplies one bus bar up to an output of 500 kva. One half of each centre trailer is supplied by train bus bar 1 (ZS 1), the other half by train bus bar 2 (ZS 2). Coaches 1, 3, 5, etc. are supplied by (ZS 1) and coaches 2, 4, 6, etc. by (ZS 2). This means that the transformers in the power units are loaded symmetrically. If one train bus bar fails, the other one takes over.

A 110 v d.c. train bus bar has been installed for the power supply of the diagnosis system, the train control, and the safety-related consumers in the event of battery failure.

The heating indicator of the air-conditioning system air-treatment unit (K 1) and the transformers are supplied from the high-voltage 1,000 v, 16⅔ Hz, the

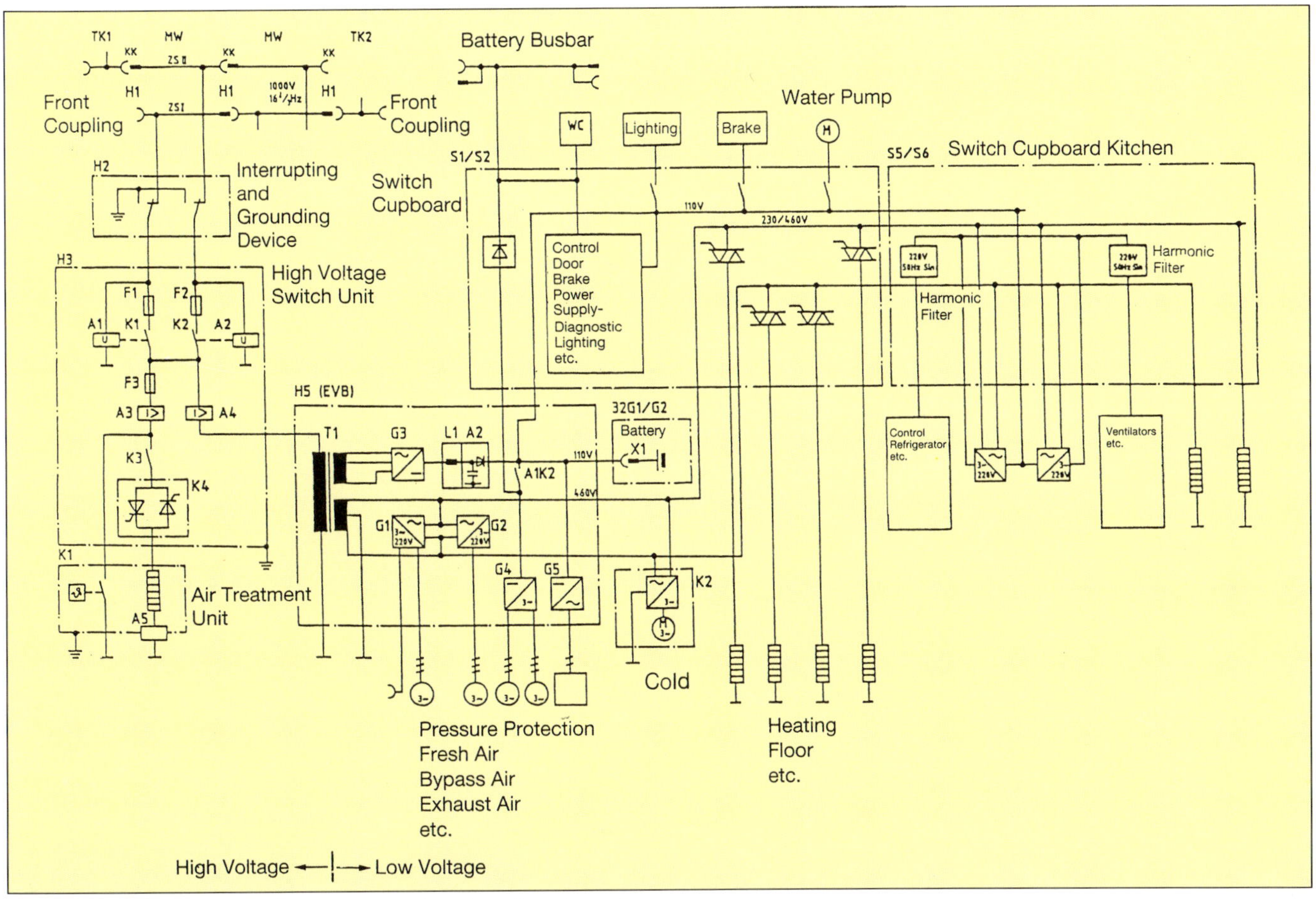

Fig. 36: Concept of the electrical power supply of the ICE centre trailers

latter with approx. 50 kva for the passenger coaches and approx. 90 kva for the restaurant car.

All the consumers of the secondary of the transformer (see Fig. 36) (T 1) are fed by the low-voltage supply.

The frequency and voltage controlled air-conditioning rotary rectifier, with an input of 460 v and an output of 26 kva and two pressure-protection rotary rectifiers for the supply and exhaust fans (G 1 and G 2), are fed directly by the transformer (T 1).

The floor heating and ancillary heating systems, for instance, are fed with 230 v. The battery charger (G 3) charges the lead batteries with 110 amp/h at 110 v.

Electrical equipment cupboards are accommodated at one end of each centre trailer and, to be accessible at the rear, can be swung out through 180°. The remaining equipment is grouped within the underfloor pan.

Fig. 37 shows the underfloor equipment configuration in the centre trailers.

Fig. 37: Arrangement of the underfloor equipment

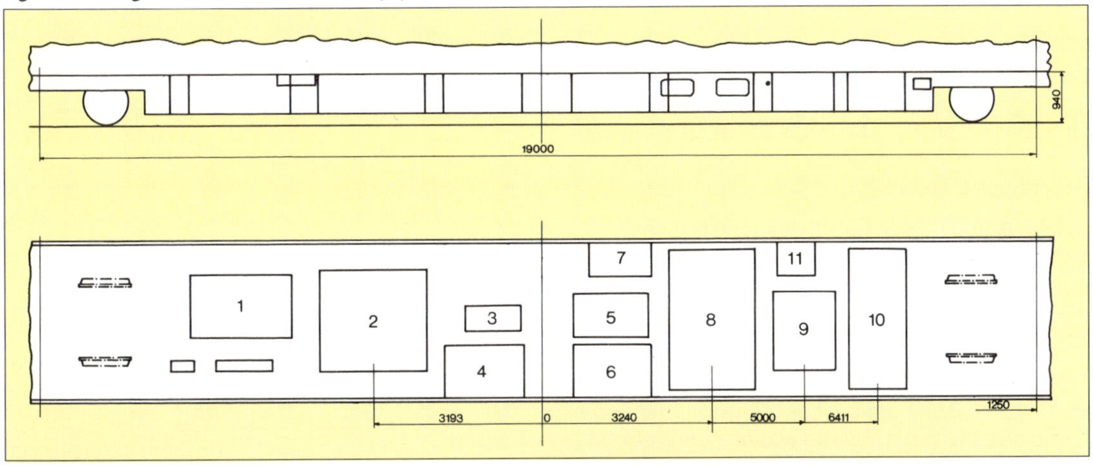

1 Airbrake Assembly
2 Electric Power Supply Unit
3 Transformer for Eddy Current Brake (Option)
4 Battery
5 Noise Damping Box for Pressure Protection Fan
6 Battery
7 Equipment Box for Eddy Current Brake (Option)
8 Compressor-Condensor Unit
9 Fresh Water Tank
10 Waste Water Tank
11 Equipment Box for Eddy Current Brake (Option)

Manfred Ackermann

Electrical power supply for the centre trailers

Operational experience gained on the prototype ICE/V and Bvmz 185 passenger coach, has enabled the design of the production version of the centre trailer to be developed further.

Because of the complexity of the problems which confronted them and the short deadlines for development and production, the electrical engineering companies engaged in building passenger coaches in Germany, formed the Konsortium Elektro- und Klimatechnik (KEK)[1]) to ensure the greatest possible collaboration and use of maximum capacity.

The major features of the new design are briefly described below and should be read in conjunction with Fig. 1.

The local power converters have been combined in the power supply block (PSB) and the control system has been incorporated into the newly developed "ZEUS" system[2]).

The cabinets for the original systems, and those for the diagnostics and passenger information (PIS) systems are force-ventilated and because of their high power density cooling has been integrated into the air conditioning system.

The contact load capacity of the couplings (H 1) has been increased because of the larger number of coaches. Also because of the high information rate necessary for train control diagnostics and passenger information, fibre-optic cable systems (LWL) have been incorporated for data transmission with redundant fibres incorporated to provide additional channel integrity.

Power for the air-conditioning system and for the central transformer (T 1) in the power supply block PSB (H 5) is supplied directly from the train bus voltage at 1000 v, 16⅔ Hz via the insulation and earthing

1) KEK: Konsortium Elektro- und Klimatechnik (ICE-MW) (Electrical and Environmental Consortium).

2) ZEUS: Zentrale Einrichtung für Überwachung und Steuerung (Central Monitoring and control device).

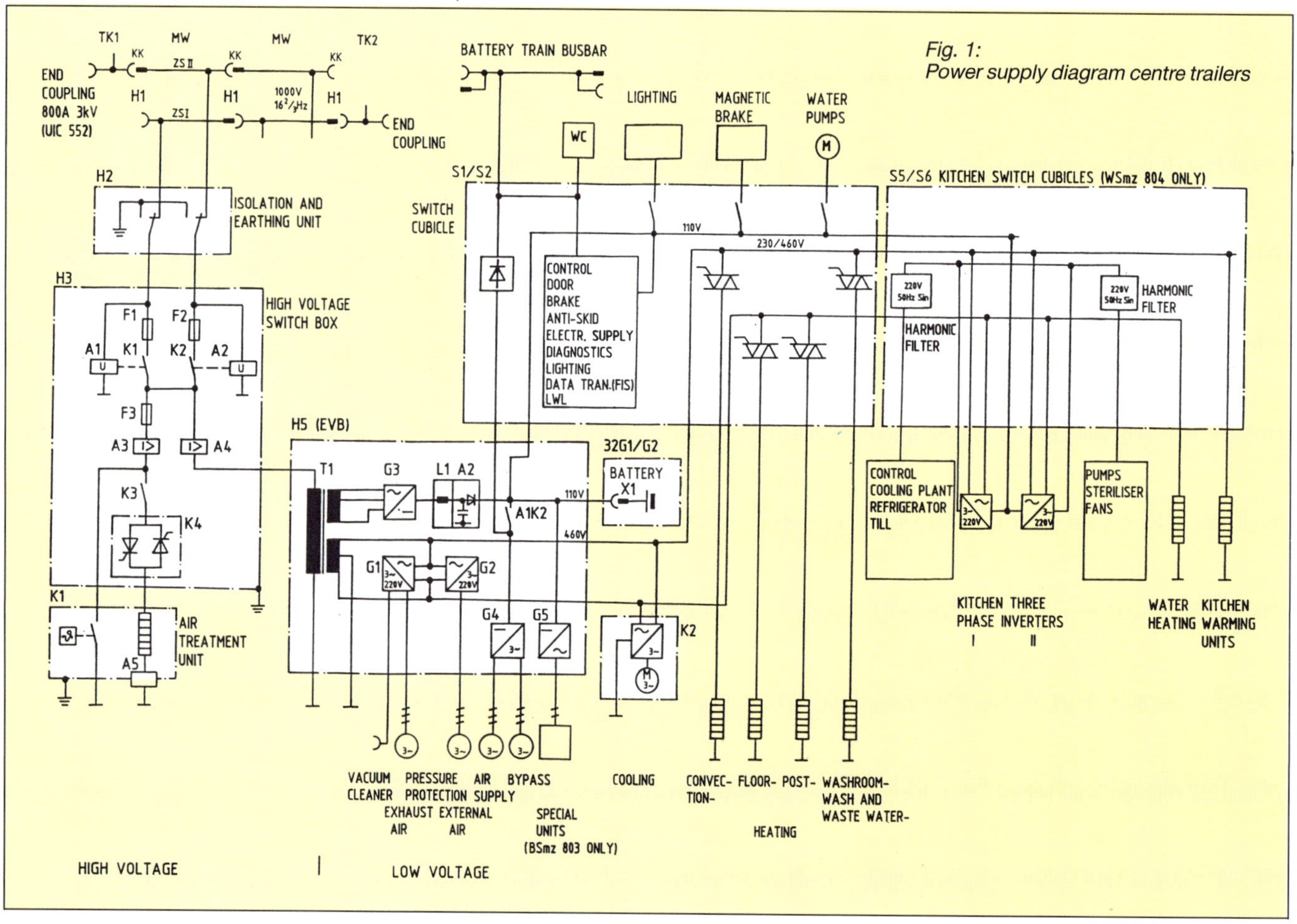

Fig. 1: Power supply diagram centre trailers

device (H 2) and the high-voltage switch cupboard (H 3).

The other supply voltages

- ▷ 230/460 v, ... 16⅔ Hz,
- ▷ 220 v, ... 50 Hz,
- ▷ 3×220 . . . 300 v, ... 68 Hz

and the d.c. voltages

- ▷ 110 v and 24 v

are produced via transformers or via single and three-phase converters from the the 110v battery voltage. The charging voltage for the battery is obtained from the transformer via the battery charger (BC) and series-connected smoothing filter.

Development of the semiconductor power and control systems has also taken place.

A clear definition of the interface between the electrical and mechanical parts has produced an optimum layout of equipment and cabling, both under and in the coach.

The introduction of the central PSB (Fig. 3), and the removal of the eddy-current brake, means that symmetrical load distribution on the two busbars is no longer possible. In summer operation for instance, as a result of the 460v feed selected for the three-phase cooler converter, the total power from the 1000v busbar is taken through the transformer. Power is therefore taken from one busbar per car, the busbars being so arranged that coaches 1, 3, 5 etc are connected through busbar 1, coaches 2, 4, 6 etc through busbar 2. This preserves symmetrical loading of the two power unit transformers.

Fig. 2: Electrical section of the ICE short coupling

System design

As in the ICE/V the electrical supply for the centre trailers is via two insulated busbars at 1000 v 16⅔ Hz (ZS 1 and ZS 2/Fig. 1). Each power unit supplies each busbar with a load of 500 kVA. The supply couplings between the vehicles have therefore been fitted with contacts further developed from the UIC 800 amp. 3 kv coupling (UIC 552) (Fig. 2).

If one busbar fails, switchover to the other is automatic. The busbar in use and the standby continue to work at half power, with the exception of the special equipment in the WSmz 804 restaurant car.

A further 110 v d.c. busbar is installed to supply the safety related loads, diagnostics and train control to cope with battery failure on a centre trailer. Fig. 1 shows the total power supply scheme with coach-specific details.

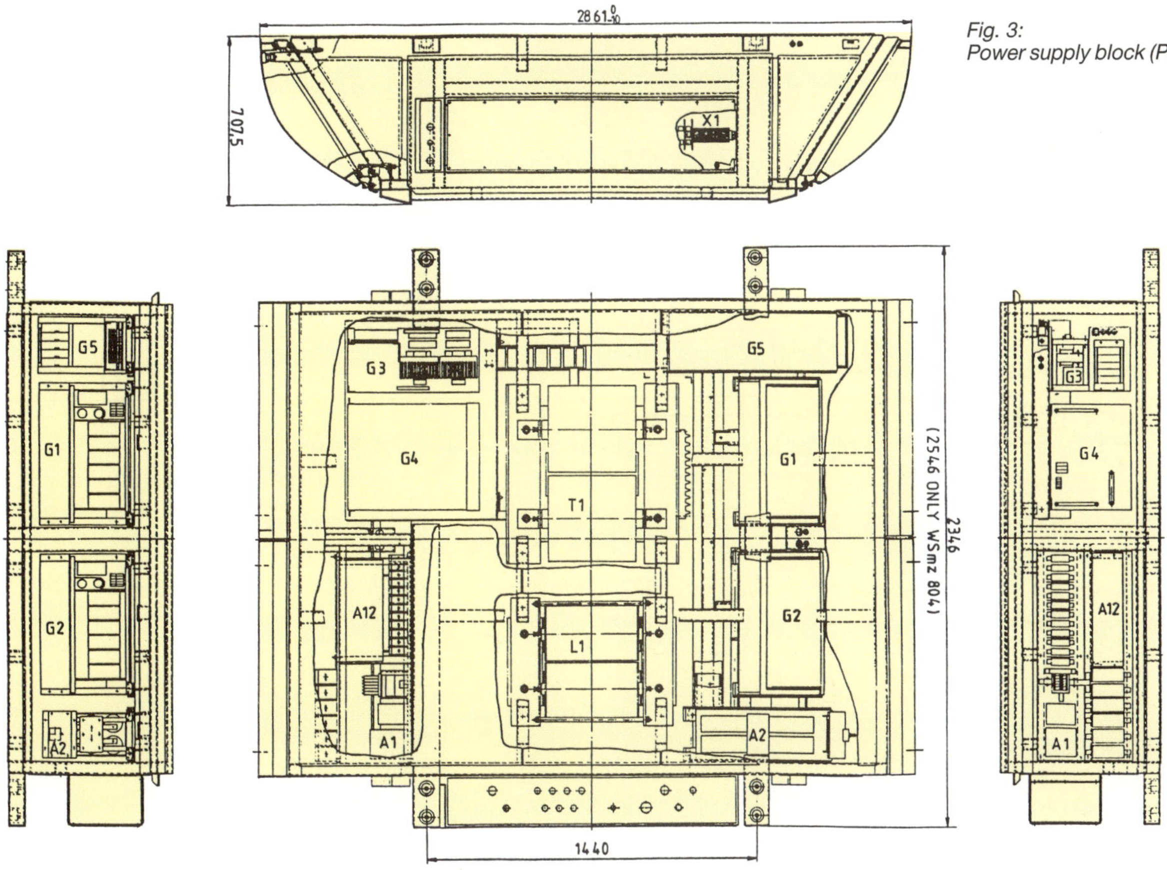

Fig. 3: Power supply block (PSB)

High voltage[3])

The heaters of the air-conditioning plant unit K 1 (LBG) and the 50 kVA transformer in Avmz 801 and Bvmz 802 are fed directly with 1000 v, 16⅔ Hz. See Fig. 3 (1 circuit 19.8 kW); (54 kVA for BSmz 803 and 90 kVA for WSmz 804).

Table 1: Connected loads

		Avmz 801 Bvmz 802	BSmz 803	WSmz 804
1000 v	Radiator of LBG (K1)	19.8 kW	19.8 kW	19.8 kW
	Transformer H5 T1 in PSB ..	50 kVA	54 kVA	90 kVA
230/460 v		32 kVA	32 kVA	71 kVA
152/190 v	for battery charging	18 kVA	22 kVA	19 kVA

Control and the switching of busbars is performed by the insulating and earthing unit (H 2). The heater is switched on by means of a newly developed 25 kW, 1000 v (Fig. 4) semiconductor, which for the extremely high duty cycles probable, can be expected to have an almost unlimited life and to operate practically maintenance-free.

3) High voltages are those above 1000 v.

Fig. 4: Semiconductor contactor

a) AEG design

b) Schaltbau design

Fig. 5: Battery charger BC

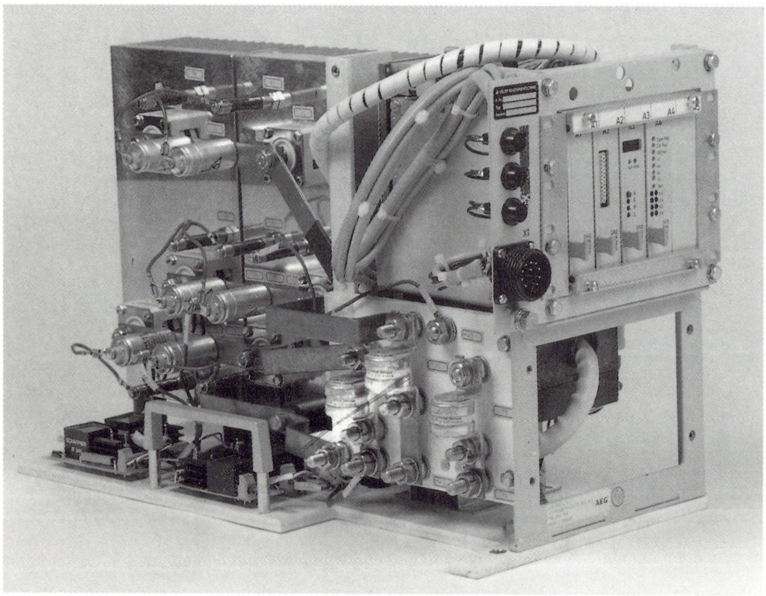

In the ICE centre trailer, the heating current monitor A 5 HzPHL 31/ICE is located in the air conditioning plant. This specially developed component – now run off a converter – was necessary because of the limited space available for pipe runs. It also includes the monitoring boards, the connection terminals for the radiator and overheating protection.

The air-cooled transformer H 5 T 1 in the PSB has two secondaries: 230 v/460 v, 16⅔ Hz for the normal loads and 152/190 v, 16⅔ Hz for battery charging. The centre tap of the 230/460 v winding is earthed (zero voltage protection mode).

Low voltage

This includes all loads from the secondary of transformer T 1 (Figs. 1 and 3).

All loads are protected on fuse panel A 1 in the PSB H 5 (Fig. 3). Circuit fuses are on switch panel S 1 (Fig. 7). With the exception of the battery ables, which are connected directly to the fuses, all cables are connected via a central terminal strip X 1 on the PSB. Battery connection is service-friendly in that it has a specially developed plug device (Fig. 7).

The following systems are fed directly from transformer T 1: – the frequency and voltage-controlled three-phase cooler inverter (input 460 v, 26 kVA), which is installed in the compressor plant (K 2); two pressure protection three-phase inverters for injection and extraction fans (G 1 and G 2) (input 460 v, 3 kVA each) in the PSB.

The following are supplied with 230 v: – floor, convection and post-heating of the air-conditioning plant and toilets, hot water and sewage heating.

The battery charger (BC – G 3) in the PSB is fed from the second secondary of the transformer and charges the lead-acid batteries via choke (L 1) and filter (A 2). These have a capacity of 110 amp/h at 110 v for the day coaches (Avmz 801, Bvmz 802, BSmz 803) and 165 Ah for the restaurant car (WSmz 804). Battery charging is to a U/I characteristic (DIN 41 70 73), the smoothing ensures an a.c. component of $0.25 \times J_5$.

Power and control systems are fitted on a common frame (Fig. 5). The microprocessor-based control section ensures problem-free matching of the various charging currents to the range of coach types.

Electrical data

Input values:

▷ Rated voltage UeN .... 152 v/190 v,
▷ Voltage range Ue .... +20%–30%,
▷ Rated frequency fN .... 16⅔ or 50 Hz.

Output values (battery)

▷ Rated voltage UB N .... 110 v,
▷ Working range at Ue N .... 120–130 v
▷ Charging current .... 110–130 amp depending on coach type,
▷ Rated charging voltage .... 120 v.

BC (G 3) and the three-phase inverter for the air supply (G 4) are, for maintenance reasons, mounted on a common draw-out unit in the PSB (H 5).

The two air pressure protection inverters (G 1 and G 2) and the filter A 2 are directly accessible from the sides of the PSB.

Equally accessible, and situated next to this draw-out unit, are the fuse panel A 1 and the measurement system MES mounted directly on it. The inductive components, transformer T 1 and choke (L 1), are arranged in the central PSB area and are accessible when the base plate is removed.

Switch cupboards

In the centre coaches (Avmz 801, Bvmz 802, BSmz 803), switch cupboards S 1 and S 2 are located at car end 1.

As in Bvmz 185, the BGT can be swung out through 180°. This feature is the result of careful development involving close collaboration between the mechanical and design partners, making use of pre-assembled panels, while maintaining service-friendly passage space.

Switch cupboard S 1 (Fig. 6) contains on its door all the necessary operating and indicating devices, including 48 small automatic switches, the BGT, control for air conditioning, door control, control system, WC, illumination, anti-skid, diagnostics, central unit (ZEUS), fan module, reserve and the MMI (man/machine interface) for ZEUS. On the 2nd level there is the triac panel for the heater control plus contactors, relays and a central terminal strip for connection to the devices in the coach.

Switch cupboard S 2, connected to switch cupboard S 1, contains in addition to the BGT for the passenger information system (PIS), the fibre-optic cable stations (LWL) and the junction box for the train control system and a terminal strip. Both switch cupboards are fully enclosed (steelplate panels) for reasons of ventilation and electromagnetic protection, and are integrated in the air conditioning system.

In S 2, air is drawn from the coach vestibule via a floor filter, led through S 1 and S 2 and extracted in the exhaust air or air circulation duct. Additional fans ensure heat dissipation from the very high-power components.

Low-voltage loads of 110 v d.c.

All control equipment, inverters, electro-magnetic rail brakes, passenger information system (PIS) and lighting are run from the 110 v d.c. on-board mains. The lighting has been newly developed in collaboration with the design and German Federal Railway (DB) departments. The reading lights, using halogen technology, have been produced after extensive measurements to provide an optimum reading environment. Table 2 shows a summary of the lighting equipment.

Where they are not supplied directly, control devices contain further subsystem converters, which generate the necessary supply voltages from the 110 v on-board mains.

An inverter fed with 110 v (DR air supply 2 kVA G 4 in the PSB) supplies the air supply fan in the air conditioning plant. It is frequency and voltage-regulated according to the state of the air-conditioning system. The water pumps are run directly via an "electronic motor" using 110 v d.c.

110 v DC train busbar

The 110 v d.c. train busbar which has been fitted has the same operating characteristics as that fitted in

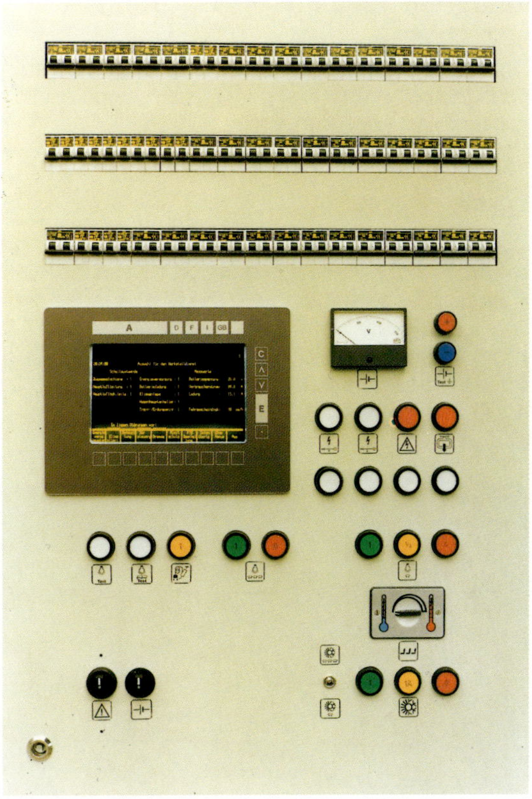

Fig. 6: Switch cubicle S 1

the ICE/V. It enables the supply to safety-relevant devices to be maintained should a coach battery fail. Suitably designed contacts are provided in the electrical section of the coupling.

Arrangement of equipment:

Underfloor space: power conversion and transmission devices and containers for fresh and waste water are built into the trough-shaped underfloor

Table 2: Lamps

	Avmz 801	Bvmz 802	BSmz 803	WSmz 804
LL 36 W	39	42	35	30
LL 18 W	5	5	12	11
Dulux 10 W	12	12	12	—
Halogen (10 W)	49	66	42	—
(20 W)	—	—	2	32

space. A segmented floor above provides convenient access for servicing.

The PSB batteries, the H 2 and H 3 units, the switch box for the magnetic brake, and the exhaust air pressure protection system are located in the centre of the coach. Cooling plant K 2 is fitted near the coach end 1.

In the coach:

Switch cupboards S 1 and S 2 are installed in the entrance area of coach end 1. Contained in the roof space of this area, is the air treatment unit (LBG K 1) with the 19.8 kW, 1000 v high-voltage heater, the pressure protection system air supply. The light switches, the guard alarm and the loudspeakers, thermostats, collected into "operating units" are provided in each compartment above the door, and in the open areas.

Light switch systems are mounted partly in the intermediate roof space and partly in the luggage rack spars as well as on the lamp rails.

Special devices:

In the BSmz 803, there is an additional inverter for the 2 kVA. 220 v, 50 Hz power supply (G 5) in the PSB. This supplies a refrigerator in the DSG – in full section, the ventilation of the PIS, switch cupboards S 3 and S 4 mounted in the centre of the trains, as well as sockets in the conference section.

Further installations are supplied by the 110 v battery supply.

In the WSmz 804 restaurant car, the PSB has a 90 kVA transformer to deal with the higher power demand. The fan inverter (G 4) is enhanced, and two additional 12 kVA three-phase inverters are installed in the underfloor space (Fig. 1).

Fig. 7: Battery plug block

In addition to switch cupboards S 1 and S 2, which are fitted opposite one another at end 1 of the coach, switch cupboards S 5 and S 6 are fitted for kitchen loads at end 2. These contain sinusoidal filters for special kitchen devices, in addition to the usual switch and indicating devices, and – like S 1 and S 2 – are fitted opposite one another. Each cupboard has its own ventilation unit.

The main air conditioning plant is accommodated in the roof space above the kitchen in the centre section. An additional plant is also included here for the air conditioning of the kitchen.

Ulrich Westerkamp and Karlheinz Althammer

ICE on record run

Introduction

After many years of intensive work, consisting of theoretical investigations, step-by-step experimental component tests in the laboratory (e.g. roller rig), as well as test runs on track carried out at the test facilities at Minden and Munich, it was possible to start the construction of the "Intercity Experimental" (ICE) in 1984. Because the previously planned test tracks at Donauried and Rheine–Freren could not be proceeded with, and the first sections of the new lines were only available from 1986 onwards, the tests were carried out on the Guetersloh–Neubeckum high-speed section of the Hanover–Cologne line, on which a German speed record of 252.2 kmph had been established as early as 1975.

The ICE record run of 317 kmph on 26th November 1985, is perhaps the most important milestone in the history of the wheel-on-rail research programme. In November 1987, another German record was established on the Hohe Warth–Burgsinn section of the new Wuerzburg–Hanover line. At 406.9 kmph (252.8 mph) a new world record was achieved on 1st May 1988, on the Wuerzburg–Mottgers section of the new Wuerzburg–Hanover line in the presence of the Ministers for Transport and for Research of the Federal Republic of Germany.

To achieve this, a number of measurements and preliminary work on the line, as well as advance tests of components were necessary, in addition to the test runs proper during the progressive high-speed trial phase. In this article the procedure adopted for this work and the most important results are briefly described.

Component tests of the production ICE

A power unit and the demonstration coach of the IC-Experimental (ICE/V) were examined in the climatic chamber at Vienna Arsenal at the beginning of 1986. The climatic equipment and the mechanical components of the ICE/V had to prove their technical efficiency. Within a temperature range of between 40° C and −25° C and at air velocities of up to 160 kmph.

It was shown that the air-conditioning plant is capable of adequately heating the relatively small, well-insulated driver's compartment at an output of only 5 kW. However because of its narrow dimensions, it is difficult to achieve uniform air distribution with central, directional air supply. The built-in floor and recess heating made an appreciable contribution to increased comfort. Micro-processor control ensured precise maintenance of the desired temperature.

Fig. 1: The ICE in the climatic chamber

Fig. 1 shows the power unit of the ICE/V in the climatic chamber in Vienna.

Because of numerous new features in the centre trailers, comprehensive metrological equipment concentrating on air and surface temperature measurement was installed there. The vehicle was examined in the heating and cooling modes according to the ORE standard programme.

The floor heating functioned perfectly and produced a comfort-enhancing temperature gradient in the vehicle, which could not have been achieved with the air supply entering through nozzles in the ceiling alone. Compared with the airliner, where this concept originated, the large window surfaces of the rail vehicle cause relatively high heat losses, which have an adverse effect on the surface temperatures and on the establishment of a uniform air circulation. In addition, some thermal bridges occurred in the aluminium extruded-profile outer wall. These were be eliminated in the production vehicle.

Together with the examination of the compartment variant for the production train, a few other possibilities for improvement were investigated in December 1986. These investigations confirmed that the air supply can be introduced more uniformly through a suitably designed perforated ceiling, so that the ambient air speeds remain within the physiological comfort range. In addition to improved heat insulation of the outer wall, a convection heater with an air shaft in front of the window most effectively increased surface temperatures to the required level. The programmed micro-processor control permits optimum regulation of the interaction between the components involved in air conditioning.

Experiments with icing-up and snowing-up of the doors were carried out at temperatures around −6° C. They confirmed that both effective sealing and trouble-free opening and closing of the doors [1] were possible in winter conditions.

Last, but not least, the brake equipment, too, had to prove its reliable function in wintry conditions. For this purpose the friction behaviour of the sinter metal linings on the solid steel brake discs of both vehicles was investigated; on the brake test rig of the driving chamber at −10° C, and in the stationary chamber the effect of low temperatures down to −20° C, as well as ice and snow on the control and on the moving parts of the brakes, especially of the eddy-current brake and the heatable sanding equipment, were successfully tested [2].

An important question within the framework of the component tests was whether the MD 522 high-speed bogie with its 2.5 m standard wheelbase would also be suitable for the production train. The 2.8 m wheelbase variant installed in the centre trailer of the ICE/V had already shown very high running qualities. In addition to the advantages of lower mass and thus reduction of its moment of inertia round the vertical axis, a shorter bogie offers more space for the installation of the structural components and appliances under the vehicle.

For the tests a production MD-522 bogie was fitted with a second hunting control linkage, with traverses for the support of a possible eddy-current brake, as well as with measuring wheelsets with tapered roller bearings. In addition, the support base of the variable side bearers was broadened by 520 mm on each side. As a carrying vehicle a production Bpmz coach was attached to the ICE/V, which served as a "high-speed locomotive".

In addition to accelerations in the coach body and in the bogie, the horizontal (Y) and the vertical (Q) forces between wheel and rail were measured with the measuring wheelsets and immediately after the run processed for 500 m-long evaluation sections, according to statistical methods of calculation. Fig. 2 show that, even with single hunting control, the limiting values according to Prud'homme are not approached up to 300 kmph. With double hunting control the Bpmz coach encountered no problems even at the maximum speed of 335 kmph attained by the ICE/V over a short distance, so that the production train will be fitted with the MD-522 bogie with 2.5 m wheelbase [3].

The Intercity Experimental is the first track vehicle equipped with a linear eddy-current brake designed to run without restriction on the lines of the German Federal Railway.

A linear eddy-current brake, like an electro-magnetic rail brake, functions independently of the value of the coefficient of friction between wheel and rail. Unlike the electro-magnetic rail brake, however, the linear eddy-current brake does not make contact with the rail; this not only makes it independent of the coefficient of friction but also non-wearing.

However during the use of the linear eddy-current brake very strong magnetic fields are generated, which may interfere with vehicle sensors near the track and with cables of the signalling and telecommunications systems, with the resulting possibility of inadmissible effects on the safety of the signalling system.

It was therefore necessary in other test and demonstration runs, by means of measurements, to establish the outline conditions under which the eddy-current brake could be used.

After the effectiveness and control of this system had already been confirmed during the first phase, the nature of the effect of the linear eddy-current brake on the permanent way had to be established.

In a combined experiment at the beginning of December 1986, the brake force of the train as well as the temperature increase and lift and depression of the rail were measured. For this purpose, the temperature was measured at nine representative sites of a rail profile and compared with a control rail. At the same time, displacement pickups indicated the vertical displacement of the rail between two sleepers under the influence of the passing wheel and of the linear eddy-current brake magnet.

After four unbraked preliminary experiments, the measuring site was passed 20 times at 10 min intervals at about 120 kmph with the linear eddy-current brake applied. With an excitation current of 200 amp, the brake force of the train was 100 kN, corresponding to 7.1 kN per individual magnet. During each braking action the temperature rose about 3.6° K. But during the cooling phase the original level was not regained, so that the rail temperature curve was

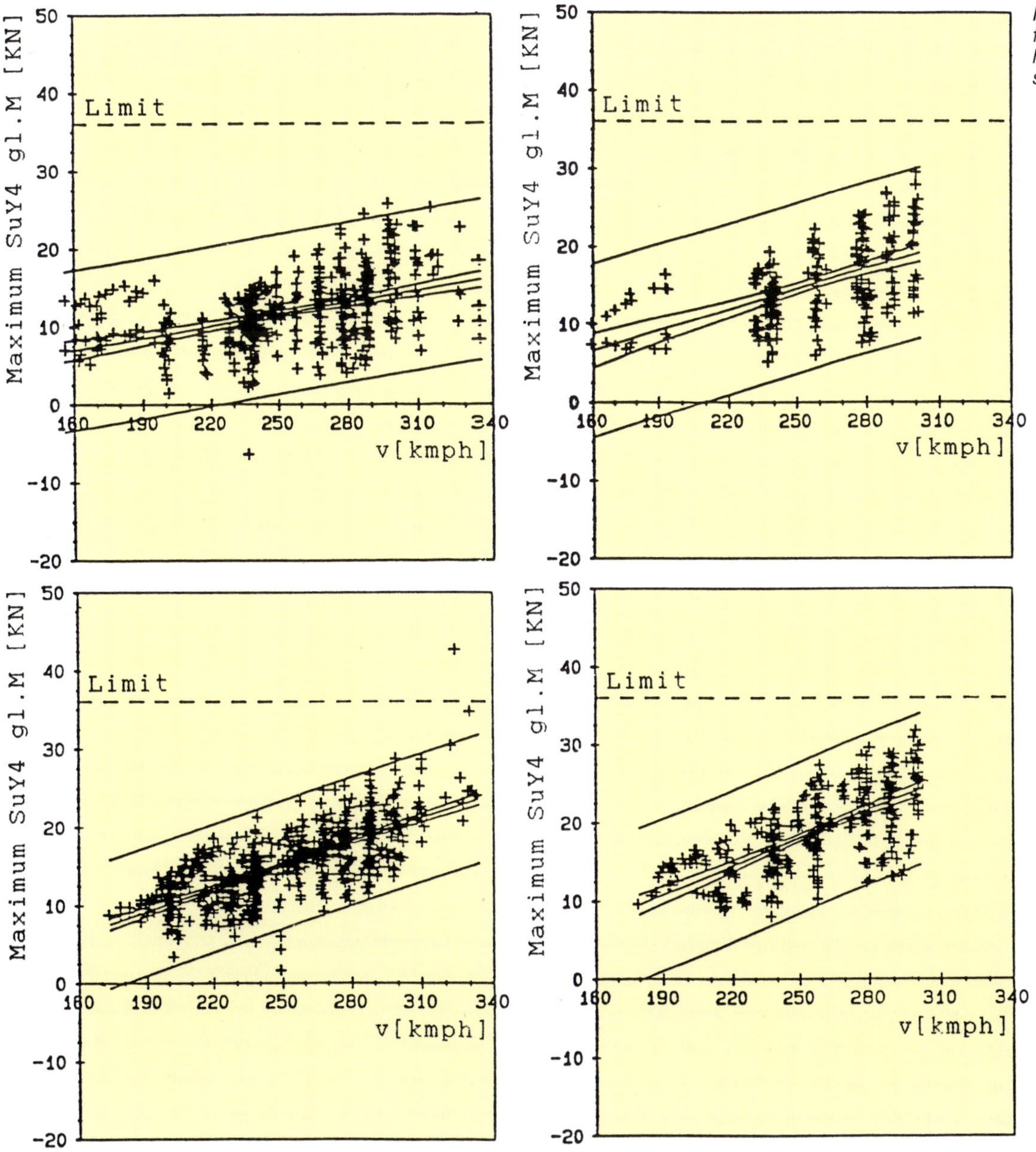

Fig. 2: ΣY-forces on the fourth axle with double hunting control – straight sections of track

jaffed. After 20 passages during this experiment, a certain final temperature was reached.

This measurement confirmed a previously carried-out simulation calculation with welcome accuracy; the energy balance, too, revealed very close correspondance between the mechanical braking energy and the quantity of heat absorbed by the rail [4].

The measurements of the rail very clearly demonstrate the depressions, which are proportional to the wheel loads and thus reflect the number of cars in the train. When compared with the unbraked pre-trial run V3 (see Fig. 3, curve 1), the braked run 3 shows only minor differences (curve 2). The precise agreement of the measured values of the trailing bogie of the second power unit is explained by the switching-off of this brake unit. Between runs 1 and 19 at 15.5° C rail temperature (curve 3), no significant difference is apparent. The rails act as hold-downs opposite the linear eddy-current brake, so that the rail cannot be lifted de facto. The same applies to switch blades [5].

Fig. 3: Depression of the rail with and without linear eddy-current brake

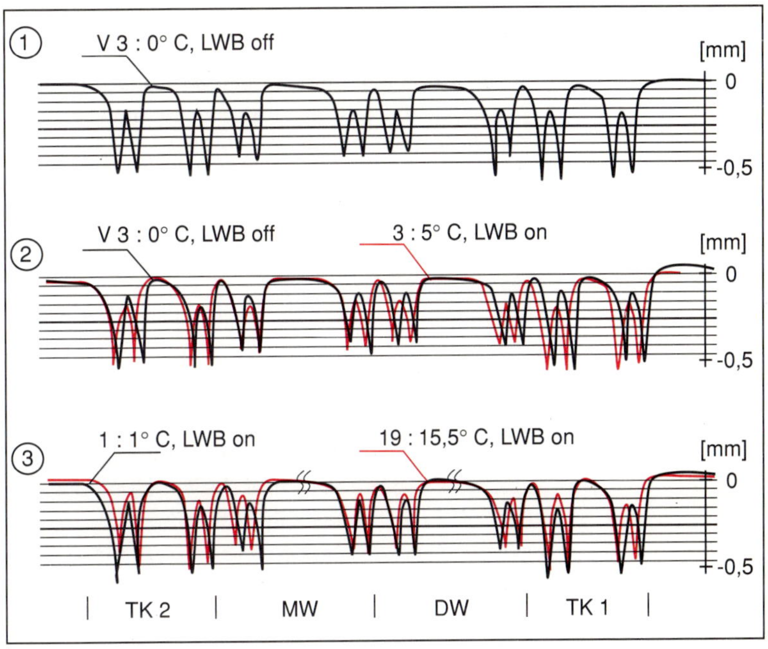

In conclusion, it can be stated that material stress of the rails caused by the rise in temperature measured during this experiment does not create a problem.

For the use of the linear eddy-current brake in a complete train of 14 units, the rail temperature will in future be measured by infra-red sensors in the leading bogie of the power unit; these sensors have already proved their suitability during temperature comparison measurements.

Preparation of the ICE and of the track for the world record run

The opening of the new line sections was keenly awaited, so that the ICE/V could be tested at a speed of 350 kmph: the speed for which it had been designed. At long last, on 15th July 1986, a 26 km section between Burgsinn and Hohe Warth north of Wuerzburg was, although not yet officially open, ready for experimental operation. After speeds of up to 280 kmph had been reached during service trials with a dynamometer train consisting of a Class 103 locomotive and three dynamometer cars, the trial runs with the ICE/V were resumed on 2nd September.

Preliminary running-dynamical calculations had shown that in the Wuerzburg–Hanover direction the ICE/V was able to attain the following speeds:

▷ as a five-car train (power unit – centre trailer – demonstration coach – demonstration coach – power unit): 348 kmph;

▷ as a four-car train (power unit – centre trailer – demonstration coach – power unit): 360 kmph;

▷ as a three-car train (power unit – centre trailer – power unit): 372 kmph.

On 17th November 1986, the five-car ICE/V, in the presence of the Minister of Transport, Dr Dollinger, attained 345 kmph, almost exactly the theoretically possible maximum speed, thereby establishing a new German record.

From this point in time it was necessary to coordinate a multitude of activities within a tight schedule:

▷ tests for further confirmation of the ICE/V specifications;

▷ tests of the components of the production ICE;

▷ tests for the assurance of the high-speed traffic, especially tests with passing passenger and freight trains;

▷ engineering works for the public opening of the line.

Alternation of tests with engineering works on a three-weekly rotation proved to be a highly practical arrangement.

The extraordinary interest in the German "Rail Star" shown by experts, politicians, and the general public at home and abroad created another priority, the organization of presentation and of demonstration runs, which had to be accommodated more and more often during the weeks of trials.

Availability of lines was somewhat improved with the opening for trial runs of another new line section, Mannheim–Graben Neudorf on the new Mannheim–Stuttgart line.

In connection with the opening of the Wuerzburg–Fulda section of new line with the change to the summer timetable on 29th May 1988, the opportunity arose of determining the real performance of the ICE. The rated value was 350 kmph + 10% according to the usual conditions governing running trials.

The ICE/V as four-car unit was meticulously serviced for this purpose, the power units were given new measuring wheelsets, the measuring technology was fully prepared and calibrated.

At the same time, a multitude of terminal and acceptance work on the line had to be coordinated and completed within the shortest possible time:

▷ acceptance runs to verify running quality;

▷ stationary permanent-way measurements;

Fig. 4: DSA 350 pantograph by Dornier with integrated insulation

▷ stationary bridge measurements;

▷ permanent-way corrections on the basis of these runs and measurements;

▷ aerodynamic measurements in tunnels, alongside the lines, and on board the vehicles;

▷ runs under load to stabilize the track bed;

▷ measuring runs for the catenary and for the pantographs;

▷ temperature measurements of converters and of traction motors;

▷ measurement of the time/speed graph;

▷ correction work on the catenary;

▷ acceptance runs for the continuous automatic train control;

▷ level measurements for the ground-to-train radio;

▷ acoustic measurements along the line and on board the vehicles;

▷ work trains for residual work on the open line and in tunnels;

▷ training runs for the drivers who on this line will work exclusively with continuous automatic train control and with cab signalling for the first time;

▷ training of the train crew with emergency exercises in tunnel.

The acceptance runs of a locomotive-hauled measuring train at 280 kmph confirmed good track bed quality, with the exception of a few irregularities; this justified expectations of runs at appreciably higher speeds.

On the open line, the catenary was the main concentration object of. It had to be prepared so that the aimed for record speed could be attained.

Previous trial runs had shown that the limiting speed for the catenary in use was about 345–350 kmph. The problem zones were, above all, the support points, particularly those in tunnel. A system was therefore devised that permitted the stationary measurement of the lift of the supports on ten predetermined sections of open line and in tunnel.

It is well known that the lift of the contact wire is affected by the longitudinal span, and by the contact wire and carrying cable tensions [6].

Because the longitudinal spans were beyond modification – the catenary was, after all, already in place – only the contact wire and the carrying cable forces could be adjusted.

Wave velocity was also important. According to calculations, as well as to drop tests of the Re 250, this is 426 kmph [7].

Because of the Doppler Effect the running speed of the train and the wave velocity should be as far apart as possible.

Theoretical considerations have shown that the wave velocity increases only when the mechanical tension σ_F [N/mm²] also increases, because the mass:spring ratio will then be affected thus, the eigenfrequency increases.

For this reason, the tension of the contact wire of 15 kN (usual in the Re 250) was increased to 21 kN, whereby the wave velocity increased to 504 kmph.

This increase in force, however, resulted in a longitudinal change (stretching) of the contact wire of up to 0.25 m so that – at the geometrically conditioned pitch of the droppers – the contact wire was no longer parallel above the track but undulating above it. As a consequence, the catenary had to be readjusted in the section where the tension of the contact wire had been increased. The following table [7] lists the different data of the Re 250 and of the experimental Re 250 catenary.

Dynamic characteristics of the Re 250 catenaries and of the Re 250 record catenary

	Re 250	Re 250 record catenary
Contact wire	CuAg Ri 120	CuAg Ri 120
Tension (kN)	15	21
Carrying cable	Bz II 70	Bz II 70
Tension (kN)	15	21
Wave velocity (kmph)	426	504
Reflection factor	0.433	0.392
Elasticity* (mm/N)	0.6/0.4	0.52/0.35
Centre of span		
Support point*	0.5/0.35	0.44/0.31
Degree of uniformity* (%)	10/6	8/6
Standard deviation (N)		
at 280 kmph	22	
at 400 kmph		26
Doppler factor		
at 250 kmph	0.260	0.337
at 400 kmph	0.032	0.115
Amplification factor		
at 250 kmph	1.7	1.2
at 400 kmph	13.5	3.4

* The second value applies to tunnel sections

For the trial runs the DSA 350 (Fig. 4) developed by Dornier was used; it had already been found successful in previous measuring runs.

As in the earlier measuring runs, the curve of the contact force between contact strips and contact wire

Fig. 5: Curve of the pressure between contact strips and contact wire at various running speeds

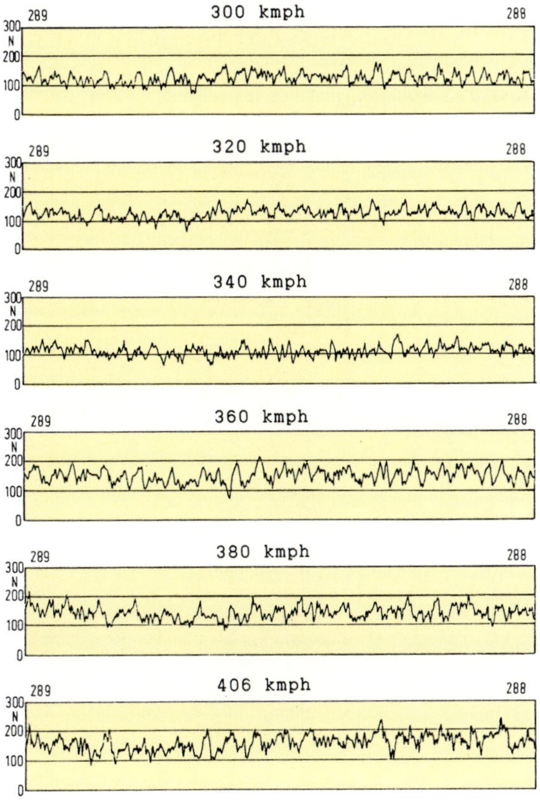

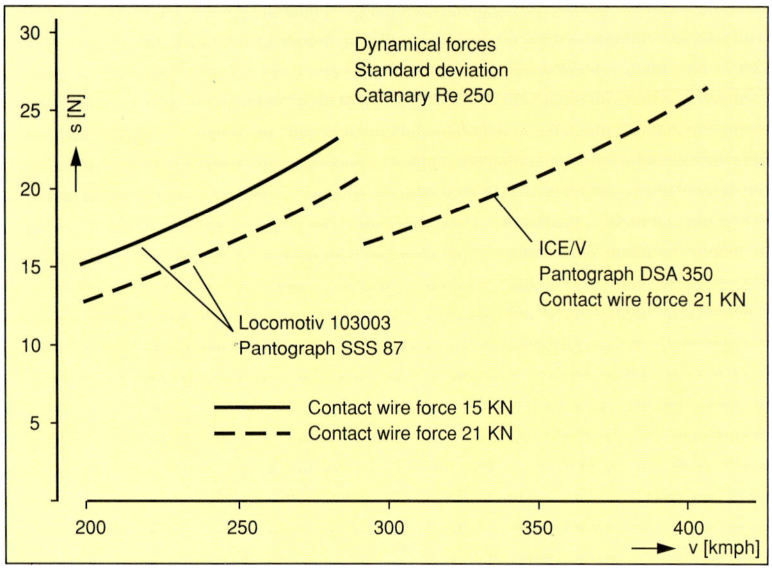

Fig. 6: Dynamical forces, impact pressure deviation with Re 250 with various contact wire tensions and various types of pantograph

was continuously measured at the various running speeds.

Fig. 5 shows the curve of these forces between km posts 289 and 288 of the Wuerzburg–Hanover line at speeds between 300 and 406 kmph.

In Fig. 6, the standard deviations of the dynamical contact forces of the SSS 87 pantograph on locomotive 103 003 and those of the DSA 350 on the ICE/V at contact wire tensions of 15 and 21 kN are entered for comparison.

It must be pointed out for the record, that strong wind forces acted on the pantograph during the high-speed runs; the maximum air velocity of 473 kmph was measured in the Einmalberg Tunnel.

Because the effective contact pressure of the pantograph at a fixed value of the air pressure in the bellows cylinder is affected by the "headwind", ie by the running speed and therefore by the reflected waves in a tunnel, the air pressure in the bellows cylinder was also controlled manually from the dynamometer car, on the basis of experience from previous trial runs. Fig. 7 shows the pneumatic scheme. Positions 4, 5, 8, 11, 12, and 13 had to be additionally installed; without which the maximum speed would hardly have been reached [8].

Fig. 7: Pneumatic circuit diagram for the manual control of the pressure in the bellows cylinder

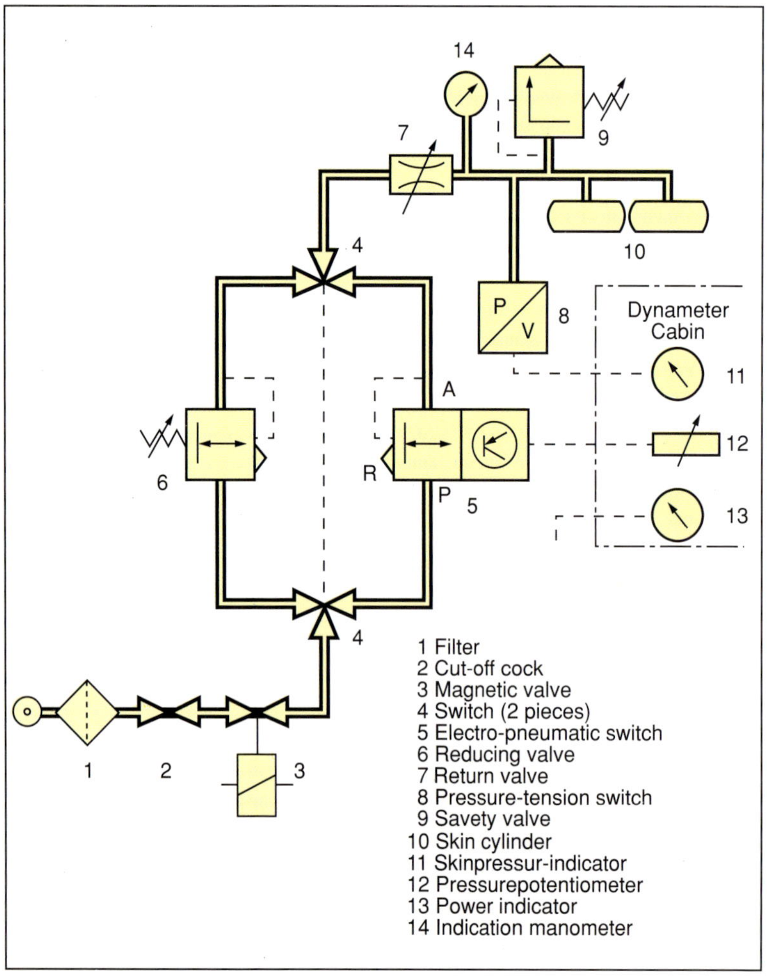

Measurement of the track and on board the vehicle in connection with the record runs

Of additional interest was the question of the stress on the foundations, the bridges, and tunnels caused by the ICE/V at high speeds in comparison with heavier loads. Particularly on the bridges it was necessary to demonstrate the real force progression in the new structural designs.

The aim of all investigations was to make data available of the static and dynamical stresses actually observed in the building structure. With such data it is possible to check the calculated statements, to assist new methods of calculation with the specification of realistic parameters, and to design and construct the regulators reliably and economically.

It was obviously impossible to examine each and every one of the numerous civil engineering structures; measurements were carried out only on typical structures and of those of special design. Of the railway bridges on the test sections, four were selected and the effects of loads of technical interest determined, such as static and dynamical loads, dependencies on the running speed, dynamical characteristics, effect of the brake forces, of the temperature, and of horizontal bearing pressures. The physically measurable variables were distorsions (dilatations, tensions), forces, displacements, vibration values, and temperatures. The measurements were carried out on important, mainly heavily stressed sites or on those that can only be calculated with difficulty, so that clear-cut statements about the internal forces, as a function of the size and position of the load, could be achieved. With these magnitudes, a comparison with calculated values was possible.

The measuring arrangement in the centre of a superstructure of the Schaippach Sinn Valley viaduct, where the experimental train reached its maximum speed, is described as an example. This structure forms a suspension bridge with individual superstructures of 42.5 m span and represents the

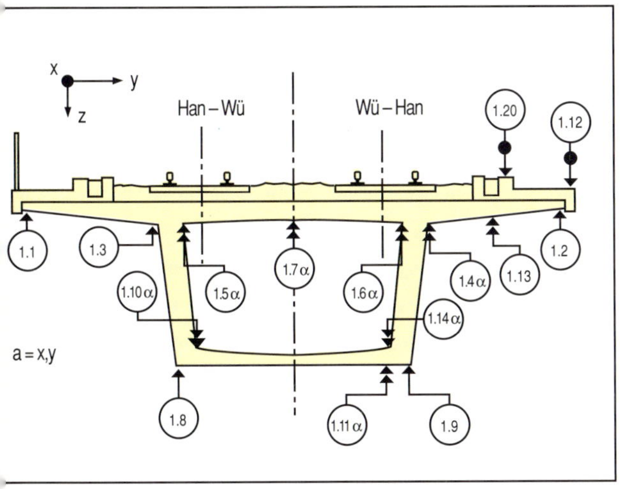

Fig. 8: Measuring site plan in the centre of a superstructure at 42.5 m span of the Schaippach Bridge (overall length 44.2 m)

.standard type of bridge on the new lines. The deformation measuring points in the longitudinal and in the transverse direction were arranged in a circle around the cross section, see Fig. 8, so that it was possible to indicate in detail the maximum or synchronous loads on the cross section. The paths of the jib vibrations (measuring points 1.12 and 1.20 in Fig. 8), with which it was possible to check the form of vibration and concrete tension, were also of interest.

Fig. 9: Dynamical effects of the load of the ICE as a function of the running speed

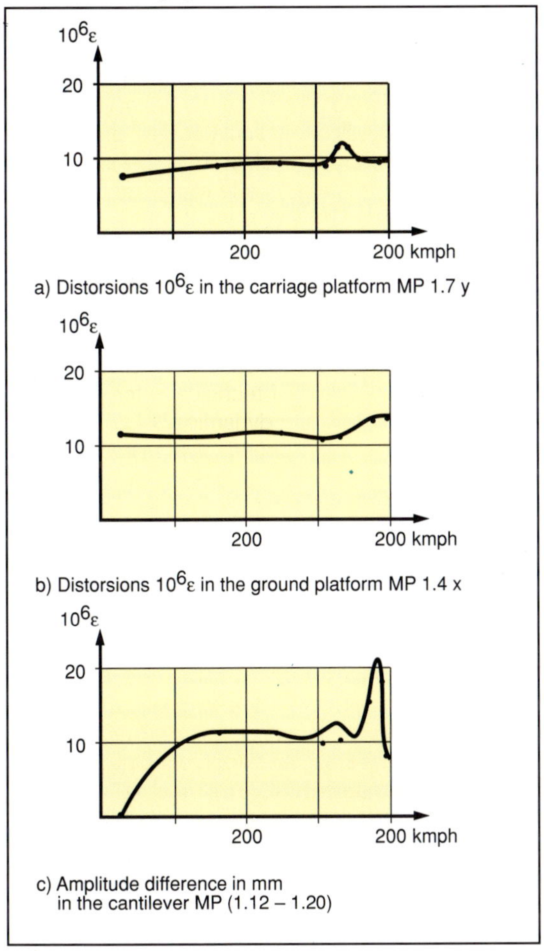

The measurement results show only a slight increase of the superstructure dynamics under the ICE/V at passing speeds of up to 400 kmph. Resonance areas are conspicuous: up to 340 kmph the eigenfrequency of the transverse flection of about 12.5 Hz is incited at the track slab, Fig. 9a; at 390 kmph the fundamental bending frequency of the vibration of the superstructure of 4.8 Hz becomes noticeable at the bed plate; it is incited by the distance of the length of the coaches, Fig. 9b; at the same speed, the fundamental bending vibration of 14 Hz of the jibs is incited by the bogie distance via the coach ends, Fig. 9c. The resonance frequencies were confirmed by oscillator experiments carried out to determine dynamical characteristic values. The torsion frequency of about 3 Hz is picked up through the entire length of the coach at passing speeds of 250 kmph; the slight increase of the measured values at 250 kmph demonstrates the minor effects of the torsion.

The static stress due to the single-track load of the ICE/V is weak, but corresponds to the lighter axle and underloads compared with those on the UIC 71 computed load diagram [9].

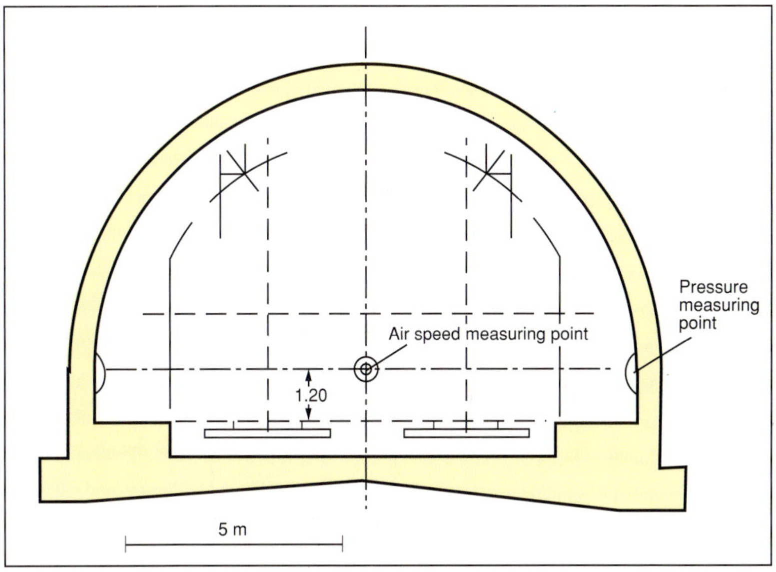

Fig. 10: Arrangement of the measuring sites in the Einmalberg Tunnel

During the progressive speed trials, the pressures and air velocities in the Einmalberg Tunnel (length 1140 m) and in the Muehlberg Tunnel (length 5928 m), as well as the internal and external pressures at various points on the train, were measured by the Munich Experimental Department.

Fig. 10 shows the arrangement of the air velocity measuring site and of the pressure measuring site in the Einmalberg Tunnel, of which there were four at 30 m and 200 m distance each from the tunnel entrances.

Fig. 11 is a diagram of the measuring arrangements in the tunnel. Because the sites of recording and the measurement cross sections are often far apart, a special measuring technique is called for. The pulse code modulating (PCM) technique has been found very reliable in measurements of this kind. Up to 16 measuring signals were converted at the measurement cross section into a 12-bit accurate digital data word, transmitted to the recording station through a

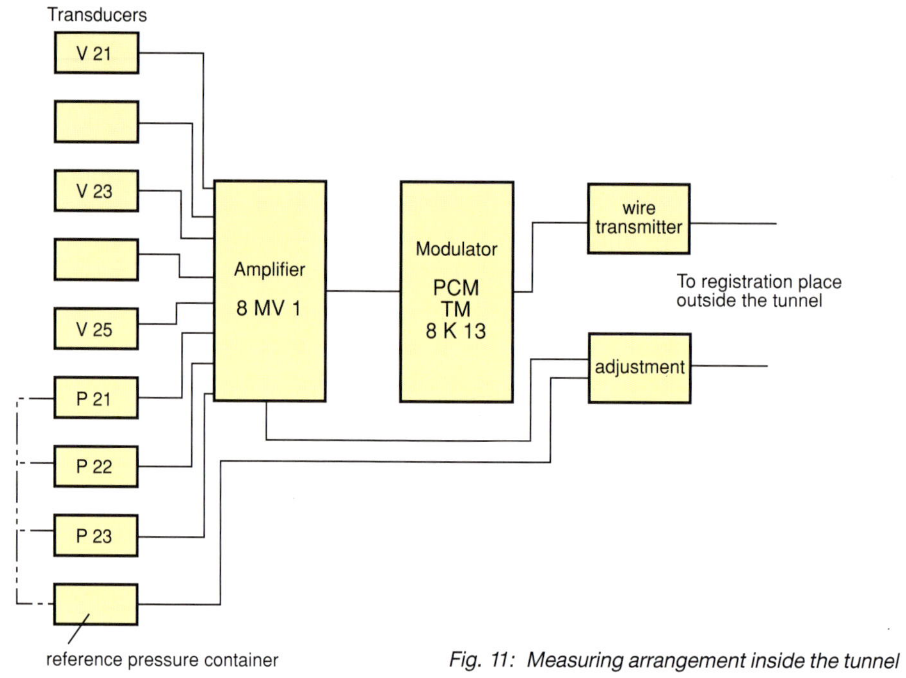

Fig. 11: Measuring arrangement inside the tunnel

Fig. 12: Recording of the measured values outside the tunnel

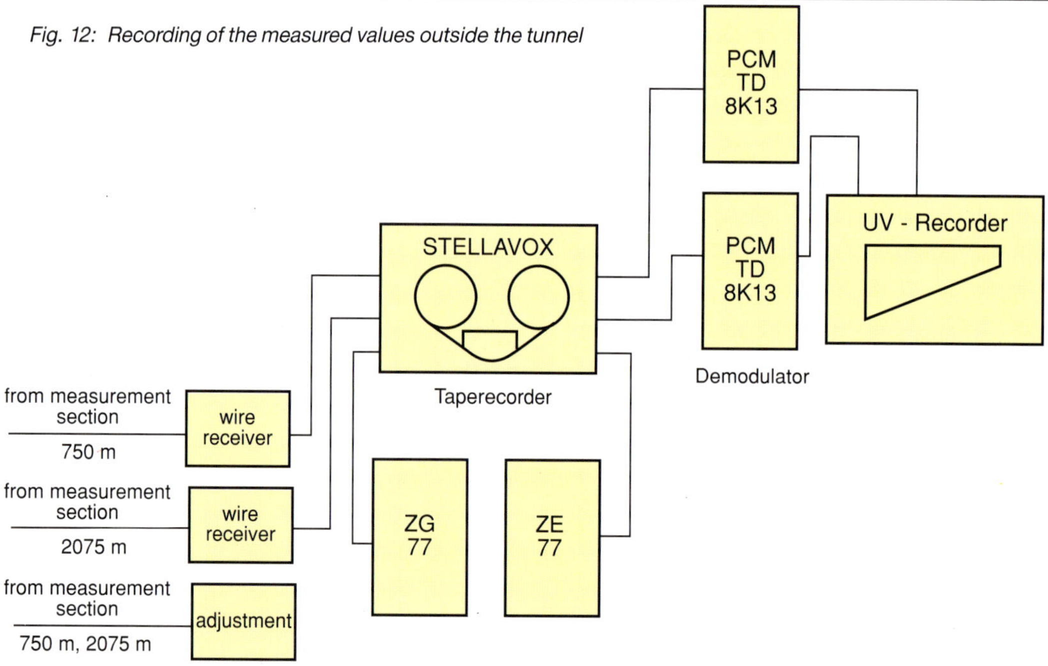

two-strand cable, which did not have to meet any special requirements.

Fig. 12 shows a diagram of the arrangement of the recording station. The measuring signals were, with the aid of the power receiver, adapted to the input impedance of the tape recorder and stored digitally on magnetic tape. They were further processed through a demodulator equipped with analogue and digital outputs:

▷ as an analogue presentation of the measuring signals on paper trace, and;

▷ as digital processing of the signals in data processing systems with suitable software.

Fig. 13 illustrates the curve of the first pressure rise, caused by the entry into the tunnel of the ICE/V power unit, at the measuring site 200 m away from the tunnel entrance as a function of the running speed. At a running speed of 250 kmph, a value of 920 Pa was recorded as the first pressure rise. As the train passed the measuring site it caused a strong pressure decrease. At 250 kmph running speed a pressure decrease of 1200 Pa, and at 400 kmph of 3120 Pa was measured.

Fig. 14 shows the air-velocity curve at track centre 1.2 m above the top of the rail, for a running-speed range of between 250 and 400 kmph. The air-velocity peak values at entry into the tunnel (E), during passage of the measuring site (V), and in the wake (N) were recorded. The dependence of the air velocities on the running speed is not as clear-cut as that of the pressures. The peak value of the air velocity was measured as 30.5 m/sec at 350 kmph in the wake, at 400 kmph, values merely between 18 and 20 m/sec were observed in the wake.

Fig. 15 plots a pressure curve and an air velocity curve at the measuring site 200 m away from the tunnel entrance. After the first pressure rise, immediately after entry into the tunnel, the rise in the compressive shear takes place. The saddle-shaped curve of the compressive shear rise is due to the differences between the cross sections of the power

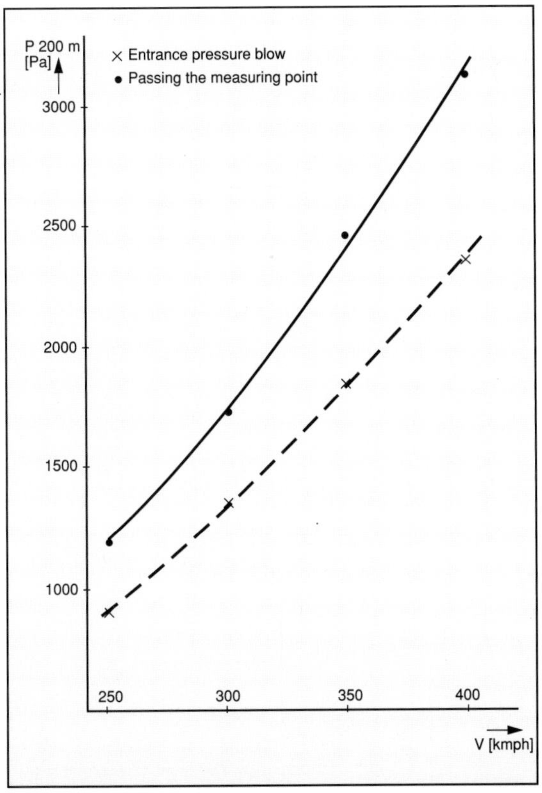

Fig. 13: Maximum pressure changes at the 200 m measuring site in the Einmalberg Tunnel as a function of the speed of the ICE

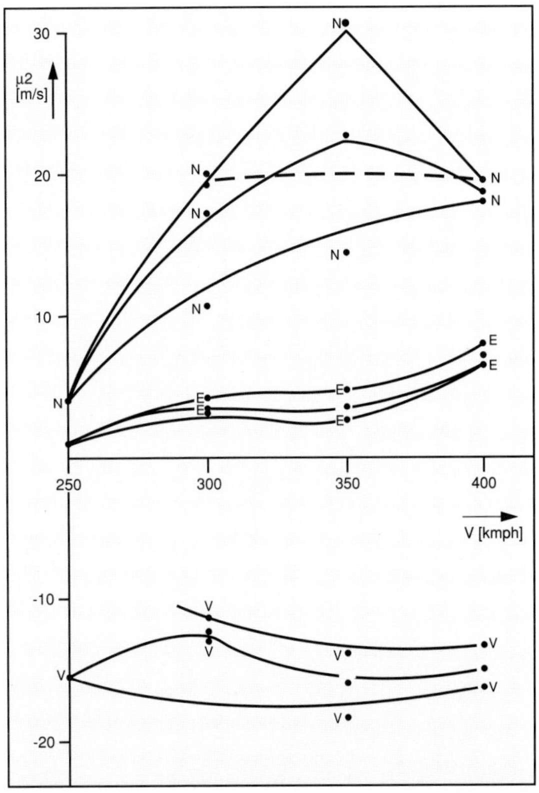

Fig. 14: Air velocities in the tunnel as a function of the train speed

E = at entry to the tunnel
V = during passing
N = in the wake
} always maximum, minimum, and mean values

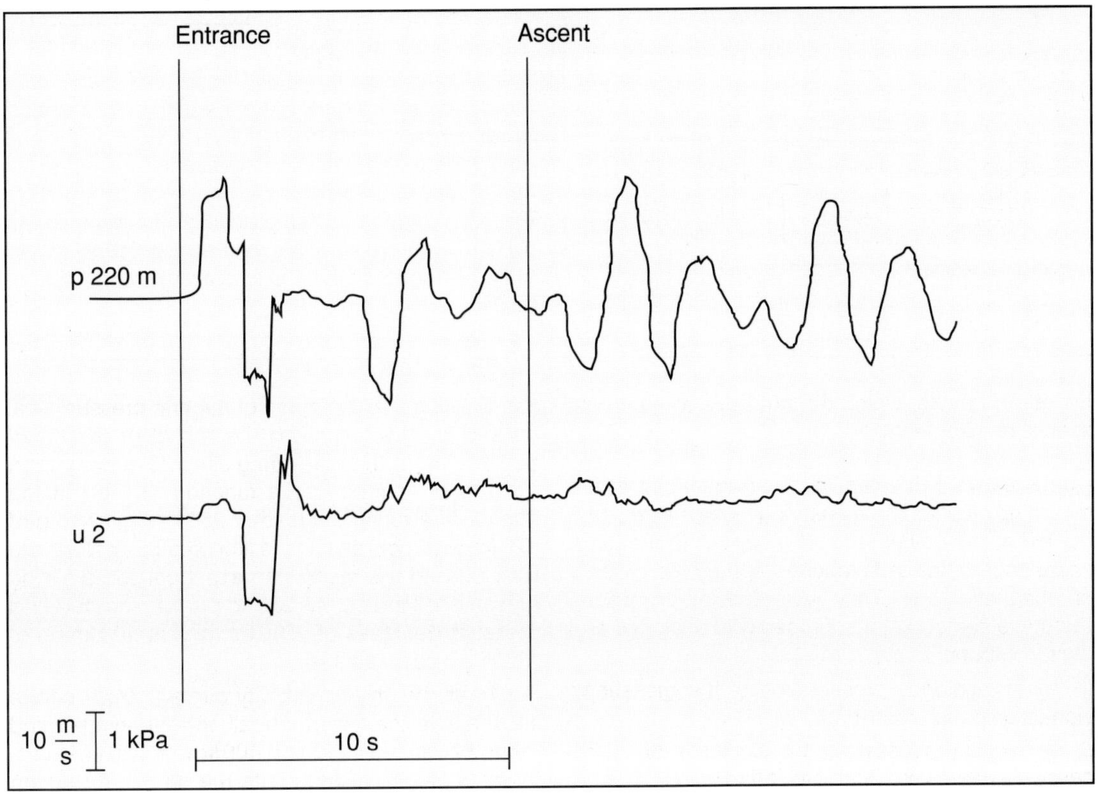

Fig. 15: Plot of a pressure curve and of the air velocity 200 m away from the entrance of the Einmalberg Tunnel

units and of the centre trailers. The entry of the end of the train into the tunnel produces a lowering of the pressure. The passage of the power unit past the measuring site causes a very steep decrease of pressure; the passage of the end of the train results in a pressure increase. The very high amplitudes of the reflections after the train has left the tunnel are remarkable. The first high amplitude reaches the same value as the pressure rise (including that of the compressive shear) generated by the entry of the train into the tunnel.

The pressure curve is very closely reflected by that of the air velocity – until the train has arrived at the measuring site. The reversal of the air flow direction as the train passes the measuring site is clearly noticeable, as is the second air flow reversal after the end of the train has passed it. Here, the peak value of the air velocity is also reached in the wake immediately after the passage of the end of the train.

After the train has left the tunnel, a relatively strong flow in the samelike direction of the train persists in the tunnel which – only the pressure – returns to its original state after a considerable lapse of time.

Fig. 16 is a diagram of the arrangement of the pressure measuring sites on the multiple unit: four external pressures and one internal one were measured. The effect of the running speed on the pressure curve is very clearly evident in the pressure rise caused by the train leaving the tunnel. Fig. 17 represents the curve of the pressure jump as a function of the running speed as the train leaves the tunnel: at 250 kmph a maximum pressure rise of 1300 Pa was

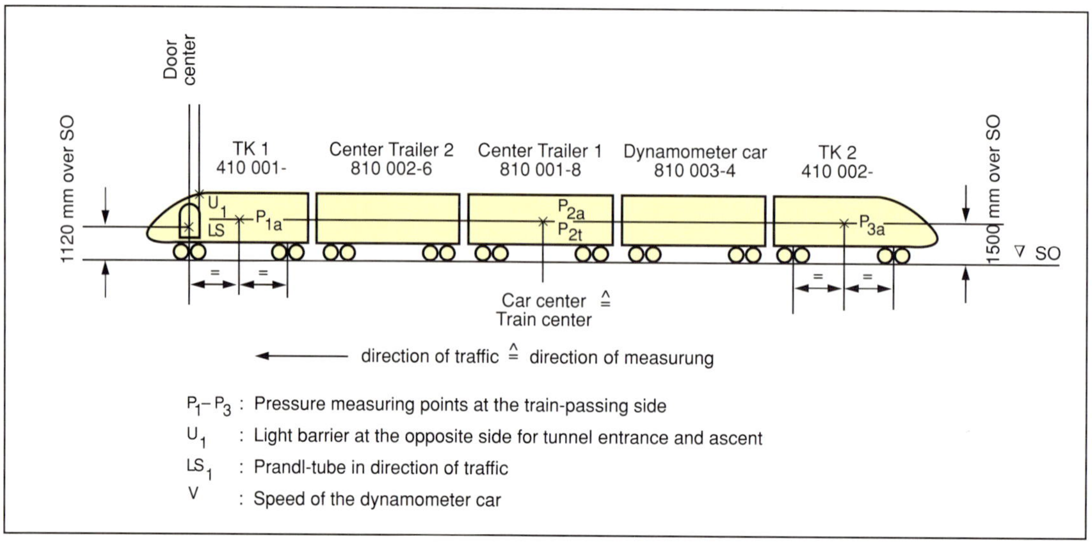

Fig. 16: Site plan for pressure measurements on the vehicle

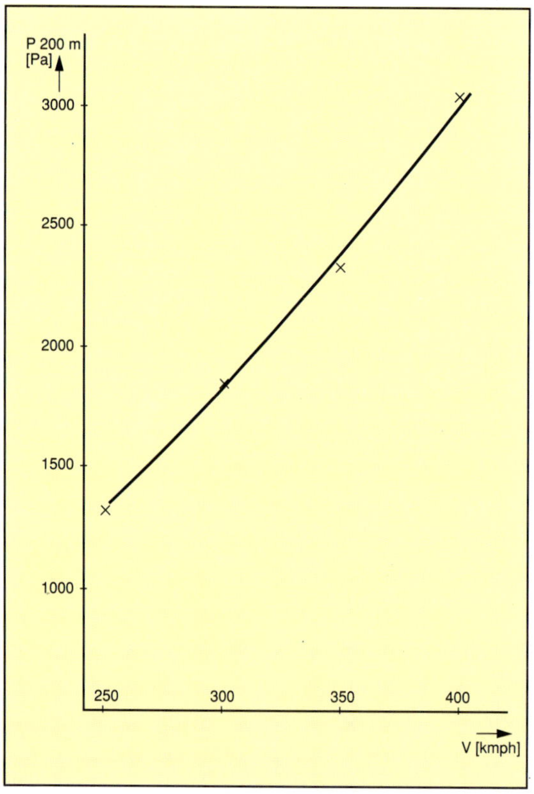

Fig. 17: Curve of the pressure increases as a function of the running speed as the train enters the tunnel

measured as the train left the Einmalberg Tunnel; this value was 3000 Pa at a running speed of 400 kmph [10].

The Munich Experimental Department also monitored the temperatures of the traction motors and of the oil cooler circuits of the converters during the trial runs, using the installed measuring sensors for this purpose. These sensors interfere with the tractive effort regulation in the drive control concept as certain temperatures are reached (traction motor 110°C+, converter 62°C+) to prevent overheating of the respective component groups.

As was to be expected, the rise in temperature at the traction motors increased with the final speed during the progressive high-speed runs. But the temperature rise of the oil cooling of the converter per progressive high-speed run depended at first approximation on the initial temperature of the cooling oil at the start of the measuring run. An example is a start-to-300 kmph at start/final temperatures of 22/34°C (Δ t = 12°C) and a start-to-404 kmph at start/final temperatures of 38/46°C (Δ t = 8.5°C). With these two runs the values of the traction motor temperatures were:

▷ 300 kmph 53/76°C (Δ t = 23°C);
▷ 404 kmph 30.5/85.5°C (Δ t = 55°C).

On the basis of these temperature measurements, the ICE/V fans were manually adjusted to maximum

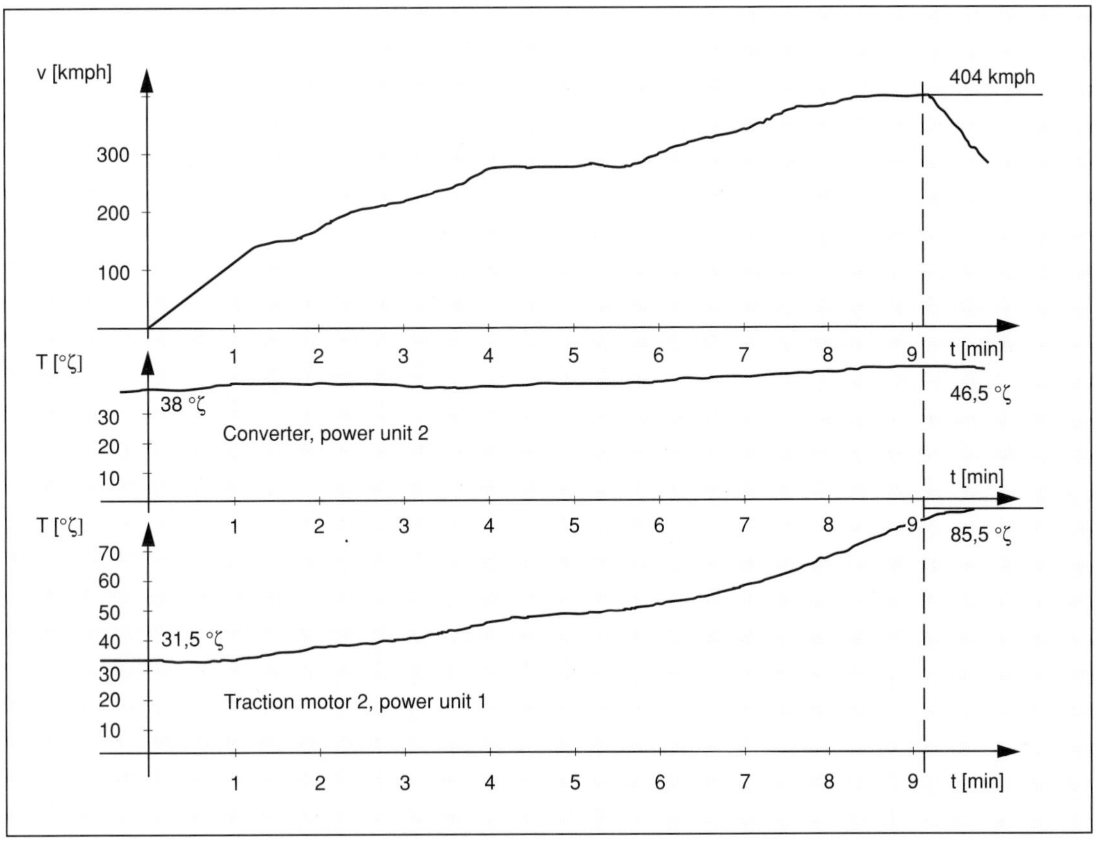

Fig. 18: Temperature curve of the converter (power unit 2) and traction motor 2 (power unit 1) at start-to-404 kmph on 29th April 1988

output from final speeds of 380 kmph and upwards, in order to lower especially the temperatures of the traction motors to between 30 and 35°C before the start of the subsequent runs. This provided a sufficiently large safety margin for the previously mentioned response values of the temperature monitors on the traction motors and on the converters.

The temperature rise at a traction motor and at the converter during the 404 kmph run on 29th April 1988 is given as an example (Fig. 18).

In addition, the time/speed curves were measured at the Munich Experimental Department, so that, to a certain extent, increased performances might be achieved in cooperation with BBC and Siemens; companies involved in the project. The time/speed curve of the 404 kmph run is illustrated in Fig. 19 [12].

The record run

In consideration of the absolute priority of a trouble-free opening of the line with the change in timetables, and the resulting need of time for any necessary final

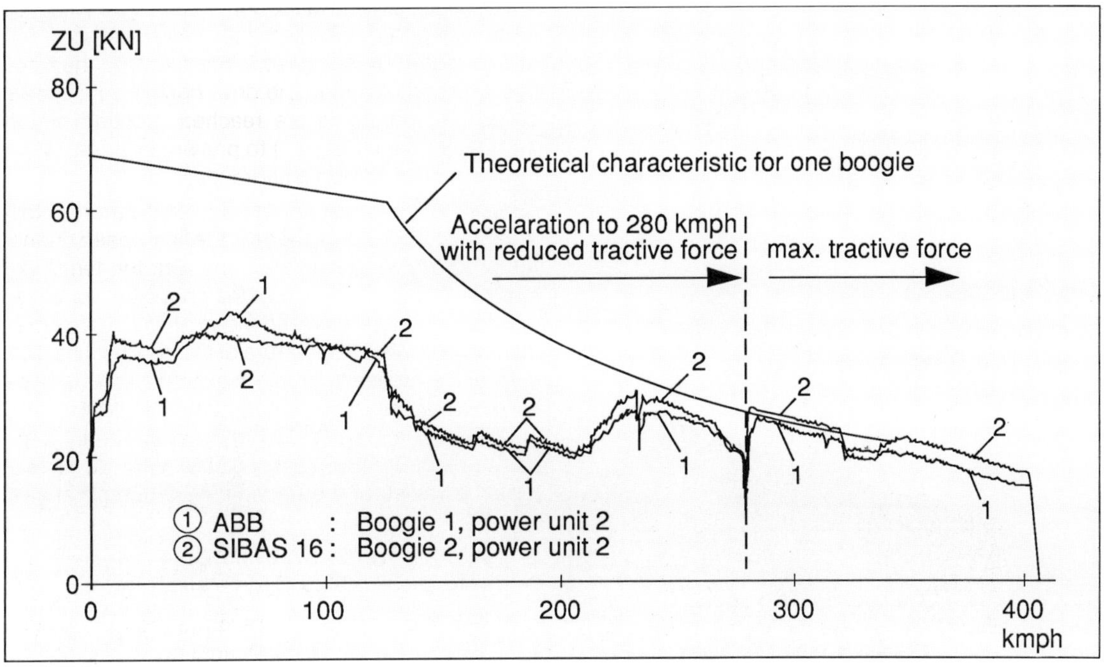

Fig. 19: Time/speed graph, start-to-404 kmph, 29th April 1988

work to be carried out, only six days were left for the progressive speed trials of the ICE/V.

The "hot" phase began on 22nd April. The first runs confirmed perfect functioning of the comprehensive measuring technology. Approximately 300 measuring points had to be covered and simultaneously evaluated to guarantee the continuous safety of the train. No fewer than ten measuring wheelsets of the Minden Experimental Department continuously recorded the forces between wheel and rail as the most important criteria for running safety.

As already described, the contact forces between pantograph and contact wire were recorded using a special measuring technique developed by the Munich Experimental Department. In addition, the pantograph function was monitored in detail with a high-speed camera. Numerous sensors monitored the temperatures of the axle bearings as well as of important components of the drive unit. Last, but not least, the drive control as well as all the functions of the brake were kept under surveillance.

The speed of the ICE/V was increased step by step, and the first target of 350 kmph + 10% = 385 kmph was reached on 27th April.

In view of the increasingly pressing question of what maximum speed the train could actually attain, limits were again imposed by the layout of the line: during acceleration a curve of r = 5300 m and of a cant of 85 mm at km post 299 could be negotiated at a maximum speed of only 360 kmph. An identical curve at km post 284 imposed a further constraint, so that a section of only 15 km length was available for speeds in excess of 360 kmph.

In the interests of safety the measured values of the running technology and of the catenary were checked immediately after each run before the speed was further increased. The drivers gradually approached the optimum driving technique and the magic barrier of 400 kmph to reach it for the first time on Thursday, 28th April 1988, and to breach it at 404 kmph on the following day.

Sunday, 1st May 1988, was the date fixed for the recordbreaking attempt, to which the management of the Federal German Railway (DB) had invited the Minister of Transport, Dr Warnke, and the Minister of Research and Technology, Dr Riesenhuber. The experimental train consisted of power unit – dynamometer car – coach for the invited guests – power unit (Fig. 20).

On a radiantly sunny day the invited guests and representatives of the press and of television, but also many railway buffs arrived at Wuerzburg to witness the expected record. Thousands of spectators lined the track, waited near the tunnel mouths and crowded onto bridges. Everyone wondered whether a new record would be set. The pressure of responsibility on the manager in charge of the trial and on the entire measuring team was intense. They had done everything necessary for its preparation; the ICE was ready to go.

Punctually at 11.00 hrs the train started to move, slowly leaving Wuerzburg Station. The guests were

Fig. 20: The ICE on the Main Valley Viaduct near Gemuenden

Fig. 21: The world record run 1st. May 1988

able to observe the track from the driver's viewpoint on monitor screens via a television camera installed in the driver's cab. At the same time they could see the km indication in the dynamometer car and the speedometer reading, which rapidly rose to 280 kmph and then remained steady, according to plan: the ICE was only to accelerate at full power from the 309 km post according to the specified running plan, so that it could negotiate the tight curve at km post 299 at exactly 360 kmph. The guests witnessed the passing of the existing German top speed of 345 kmph, the experts exchanged brief glances as the ICE/V negotiated the curve exactly as planned: they felt assured of success. The passengers and, later, millions of television viewers at home, watched the smoothness of the train's running with amazement. The figures on the luminous display climbed further to 370, 380, 390, and approached the 400 kmph mark. They reached it at 11.11 hrs; but the ICE/V accelerated further, exceeded the best test performance and a few seconds later at 406.9 kmph (Fig. 21) established another world record for rail vehicles.

Fig. 22 shows the graph of the world record run on 1st May 1988.

With this record run, the Federal German Railway (DB) and German industry have made a convincing contribution towards the efficiency of the wheel-on-rail system in competition with other transport technologies.

Fig. 22: Graphic record of the world record run on 1st May 1988 and line profile

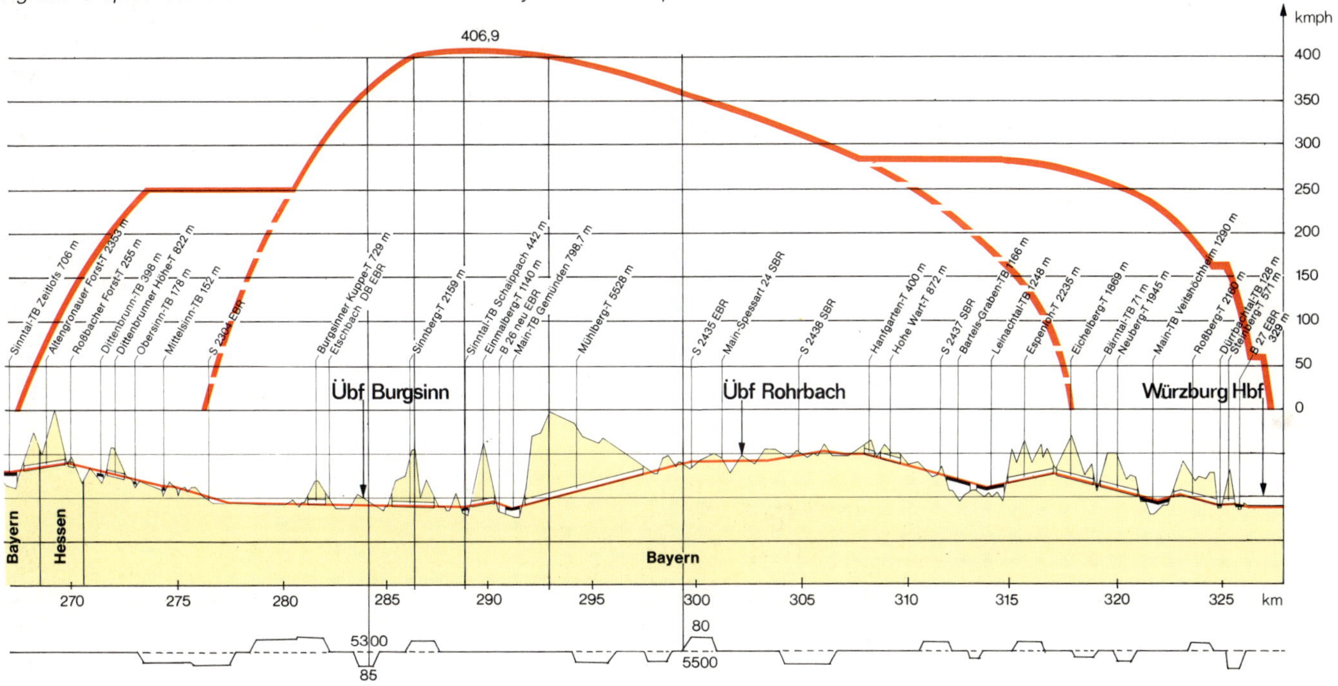

Sources of illustrations:

ABB Verkehrstechnik GmbH (3), AEG Westinghouse Transportsysteme GmbH (18), Arge ICE-Triebkopf (2), Blaupunkt (1), Deutsche Bundesbahn (8), Deutsche Bundesbahn, Film- und Bildstelle (16), Deutsche Bundesbahn, Karten- und Luftbildstelle (1), Bundesbahndirektion Köln (2), Bundesbahndirektion Nürnberg (22), Bundesbahn-Versuchsanstalt Minden (Westf) (3), Bundesbahn-Versuchsanstalt München (15), Bundesbahn-Zentralamt München (15), Deutsche Forschungs- und Versuchsanstalt für Luft- und Raumfahrt (6), Dornier GmbH Deutsche Aerospace (4), Dyckerhof & Widmann AG (3), Forschungs- und Versuchsanstalt Wien-Arsenal (1), Industrieanlagen-Betriebsgesellschaft mbH (6), KEK Konsortium Elektro- und Klimatechnik (3), Krauss-Maffei Verkehrstechnik GmbH (1), Krupp Maschinentechnik GmbH (11), Knorr-Bremse AG (10), Linke-Hofmann-Busch GmbH (3), MAN-Technologie AG (10), Messerschmidt-Bölkow-Blohm GmbH (16), Projektgruppe H/W Süd (1), Rossberg (1), Ruckhäberle (1), Schaltbau Gesellschaft mbH (2), aus A. Scherl „Ein neues Schnellbahnsystem" (4), SEL AG (18), Siemens AG Verkehrstechnik (18), Strabag Bau AG (1), Thyssen Industrie AG Henschel (7), Technische Universität Berlin (1), Verkehrsmuseum Nürnberg (2), Waggon Union (4), Züblin (2).